SpringerBriefs in Archaeology

SpringerBriefs in Underwater Archaeology

Series Editor
Annalies Corbin, Columbus, OH, USA

Books in the SpringerBriefs in Underwater Archaeology series, published in cooperation with the Advisory Council on Underwater Archaeology (ACUA), will address critical contemporary problems and illustrate exemplary work in maritime cultural heritage in countries around the globe. The series will take a broad view of the theory, concepts, issues and complexities associated with the management and protection of maritime cultural heritage sites in the 21st century. The book series will be concerned with:

- submerged cultural resources ranging in age from the most ancient human history through the modern era
- maritime cultural heritage, understood as not just archaeological sites but also the stakeholders who inevitably constitute diverse and multi-scalar sectors of society
- management that transcends adherence simply to regulation but that looks forward to the needs of the future
- emerging technologies that will transform the ways in which archaeological research is conducted and archaeological resources, both sites and landscapes, are understood and interpreted
- site stewardship and dissemination of information that serves the needs of the archaeological field and the public at large.

All volumes will be stringently peer-reviewed, first by the ACUA series editor and associate editors, and upon passing that level, two outside reviewers familiar and qualified to make constructive comments regarding the manuscript.

Charlotte Jarvis
Editor

Threats to Our Ocean Heritage: Bottom Trawling

Editor
Charlotte Jarvis
The Ocean Foundation
Washington, DC, USA

ISSN 1861-6623   ISSN 2192-4910 (electronic)
SpringerBriefs in Archaeology
ISSN 2625-2562   ISSN 2625-2570 (electronic)
SpringerBriefs in Underwater Archaeology
ISBN 978-3-031-57952-3   ISBN 978-3-031-57953-0 (eBook)
https://doi.org/10.1007/978-3-031-57953-0

This work was supported by Ocean Foundation

© The Editor(s) (if applicable) and The Author(s) 2024. This book is an open access publication.

Open Access This book is licensed under the terms of the Creative Commons Attribution 4.0 International License (http://creativecommons.org/licenses/by/4.0/), which permits use, sharing, adaptation, distribution and reproduction in any medium or format, as long as you give appropriate credit to the original author(s) and the source, provide a link to the Creative Commons license and indicate if changes were made.

The images or other third party material in this book are included in the book's Creative Commons license, unless indicated otherwise in a credit line to the material. If material is not included in the book's Creative Commons license and your intended use is not permitted by statutory regulation or exceeds the permitted use, you will need to obtain permission directly from the copyright holder.

The use of general descriptive names, registered names, trademarks, service marks, etc. in this publication does not imply, even in the absence of a specific statement, that such names are exempt from the relevant protective laws and regulations and therefore free for general use.

The publisher, the authors and the editors are safe to assume that the advice and information in this book are believed to be true and accurate at the date of publication. Neither the publisher nor the authors or the editors give a warranty, expressed or implied, with respect to the material contained herein or for any errors or omissions that may have been made. The publisher remains neutral with regard to jurisdictional claims in published maps and institutional affiliations.

This Springer imprint is published by the registered company Springer Nature Switzerland AG
The registered company address is: Gewerbestrasse 11, 6330 Cham, Switzerland

Paper in this product is recyclable.

Foreword: The Importance of Integrating Cultural Heritage into the UN Ocean Decade

The broadest aim of the UN Decade of Ocean Science for Sustainable Development 2021-2030 is to build scientific capacity and generate knowledge. The initiative promotes a common framework for supporting stakeholders to study and assess the health of the world's ocean. By design, the Ocean Decade fully recognises that there needs to be a 'paradigm shift' in how ocean science is obtained, and knowledge is generated (IOC/UNESCO, 2020a, p. 6).

The relevance of cultural heritage to the Ocean Decade is due to the direct link the initiative makes between ocean science and sustainable development, and ultimately between science and tangible changes in the world's conditions. The Ocean Decade is therefore also directly linked to the United Nations' (UN) 2030 Sustainable Development Agenda and Sustainable Development Goals (SDGs). The goal that expressly addresses the marine environment, SDG 14 'Life below water', advocates for the conservation and sustainable use of 'the oceans, seas and marine resources for sustainable development',[1] but there is no specific SDG that addresses culture or cultural heritage, although it is a target of SDG 11. The ocean contains interwoven resources, both natural and cultural – tangible and intangible, that we want to conserve and pass onto future generations: our Ocean Heritage. Knowledge generated from Ocean Heritage data about materials and societies of the past has a significant role in delivering many of the SDGs,[2] and all the more so with SDG 14.

Public interest in both tangible and intangible cultural heritage is considerable, and it is often an integral characteristic of what people seek out and enjoy at the coast and under water. Cultural heritage already makes a significant contribution to the environmental, social, and economic pillars of sustainable development – at least in some parts of the world and in some specific marine environments. Unfortunately, this is under-recognised and relatively little has been done to support or enhance the contribution already made by cultural heritage globally. The Ocean

---

[1] https://sdgs.un.org/goals/goal14

[2] Noting here the relevance of UNESCO's Thematic Indicators for Culture in the 2030 Agenda to the Decade: https://whc.unesco.org/en/culture2030indicators

Decade can provide an opportunity to ensure that insights from studying the past play a full role in shaping the transformative ocean sciences. The Ocean Decade provides us with an opportune moment to reflect: How is our ocean investigated? How can the historical dimension of people's relationship with the sea be better integrated within ocean science? How can we ensure that cultural heritage is perceived as integral natural and cultural heritage within maritime landscapes and seascapes – Ocean Heritage? How can we ensure that this Ocean Heritage is holistically considered, protected, and managed?

The UN Ocean Decade's Implementation Plan is built up from Actions which are initiated and undertaken by stakeholders and contribute to the Decade's Objectives and Challenges, ultimately helping to achieve seven Societal Outcomes:

1. **A clean ocean** where sources of pollution are identified and reduced or removed.
2. **A healthy and resilient ocean** where marine ecosystems are understood, protected, restored, and managed.
3. **A productive ocean** supporting sustainable food supply and a sustainable ocean economy.
4. **A predicted ocean** where society understands and can respond to changing ocean conditions.
5. **A safe ocean** where life and livelihoods are protected from ocean-related hazards.
6. **An accessible ocean** with open and equitable access to data, information and technology, and innovation.
7. **An inspiring and engaging ocean** where society understands and values the ocean in relation to human wellbeing and sustainable development (IOC/UNESCO, 2020b).

Clearly, the emphasis of the Ocean Decade was conceived to focus on the contribution and data of the physical oceanographic sciences such as hydrography, biological, chemical, and geological oceanography. The fields of studying and documenting the tangible remains of cultural Ocean Heritage (maritime archaeology and archaeological oceanography) are usually perceived as minor partners to or on the receiving end of data from these other disciplines. This should not be the case.

Human interaction with the historic environment, embodied in intangible traditional knowledge, is essential to understanding the state and health of our present ocean and to help forecast change, in particular related to coastal fisheries. Tangible cultural heritage, embodied by coastal sites as well as shallow- and deep-water sites like submerged settlements and shipwrecks, can help us understand how marine ecosystems achieved their present form and to identify the pressures upon them. Studies of cultural heritage can provide historical datasets to help us gauge patterns regarding pollution, impacts from climate change, and other short- and long-term hazards such as deep-sea trawling and ocean acidification (Trakadas, 2022). This argument can be perhaps outlined more clearly by framing it within the Ocean Decade's Outcomes:

1. **A clean ocean:** Cultural heritage can contribute to a clean ocean by enabling better understanding of the extent and risks of legacy pollution from shipwrecks, mining waste, and land-based sources. A clean ocean is also important for the long-term preservation of cultural heritage.
2. **A healthy and resilient ocean:** Culture heritage is fundamental to understanding how many coastal and marine ecosystems achieved their present form and to understanding the pressures upon them. Cultural heritage can be an important component of marine ecosystems.
3. **A predicted ocean:** Understanding the 'Ocean Past' (human interaction with the historic environment) is essential to understanding our ocean at present and to forecasting change and its implications for human well-being and livelihoods.
4. **A safe ocean:** Cultural heritage informs the understanding of coastal inhabitation and intervention in the past and present, including the impact of previous catastrophes, to identify risks, present examples of human adaptations, and encourage resilience.
5. **A sustainably harvested and productive ocean:** Cultural heritage is a major contributor to the Blue Economy, especially through recreation and tourism, and increasing productivity should enhance, not damage, irreplaceable cultural heritage. This volume specifically focuses on how trawling damages and can even remove cultural heritage on the seabed – it is not only unsafe for fisherman, but detrimental to our shared Ocean Heritage (Fig. 1).

Fig. 1 The wreck of a barge, likely damaged by bottom trawling, located in northern Qatari waters recording during the first week of the Qatar MAP project, May 2017. Diving was undertaken from the Qatar University research vessel *RV Janan*. Sketchfab image Courtesy Maritime Archaeology Trust, UK

6. **A transparent and accessible ocean:** Information about cultural heritage is also essential to understanding the past, present, and future of humanity's relationship with the seas and oceans.
7. **An inspiring and engaging ocean:** Information about cultural heritage is fascinating to the public and enables engagement with many topics of Ocean Literacy (modified from Trakadas et al., 2019).

In order to achieve the desired Outcomes of the Ocean Decade, each require a historical perspective on human-ocean interactions. Therefore, the Decade presents an opportunity to increase the contribution of cultural heritage to transformative ocean research and to plan for its present and future management. I hope that by 2030 – at the end of the Ocean Decade – we have arrived at a paradigm shift, where cultural heritage is considered an integral ocean science data set.

Throughout the Ocean Decade, a direct link is emphasised between science and measurable changes in the world's conditions. Halting the decline of the marine environments depends on people: hence, understanding how people behave and have behaved with respect to the sea is critically important. The wealth of knowledge generated from Ocean Heritage data about past materials and societies' interactions with the sea can and will play a significant role in delivering SDG 14. Our contention is not just 'how can cultural heritage help deliver the Ocean Decade, but without cultural heritage, how can you deliver the Decade' (Trakadas et al., 2019, p. 163)?

The Ocean Decade Heritage Network Athena Trakadas
Copenhagen, Denmark

References

Intergovernmental Oceanographic Commission (IOC)/UNESCO. (2020a). *Implementation plan summary*. Intergovernmental Oceanographic Commission (IOC)/UNESCO. https://www.oceandecadeheritage.org/wp-content/uploads/2020/10/687-20-IOC-Decade-Implementation-Plan-Summary-compressed_1597065320-1.pdf

Intergovernmental Oceanographic Commission (IOC)/UNESCO. (2020b). *UN decade of ocean science for sustainable development (2021-30) implementation plan version 2.0*. Intergovernmental Oceanographic Commission (IOC)/UNESCO. https://www.oceandecadeheritage.org/wp-content/uploads/2020/09/Ocean-Decade-Implementation-Plan-Version-2-0-compressed_1599176045.pdf

Trakadas, A. (2022). The cultural heritage framework programme: Ensuring a place for cultural heritage's contribution to the UN decade of ocean science. *Marine Technology Society Journal, 56*(3), 110–111. https://doi.org/10.4031/MTSJ.56.3.29

Trakadas, A., Firth, A., Gregory, D., Elkin, D., Guerin, U., Henderson, J., Kimura, J., Scott-Ireton, D., Shashoua, Y., Underwood, C., & Viduka, A. (2019). The ocean decade heritage network: Integrating cultural heritage within the UN decade of ocean science 2021–2030. *Journal of Maritime Archaeology, 14*(2), 153–165. https://doi.org/10.1007/s11457-019-09241-0

Acknowledgements

I first encountered the threat to the ocean from bottom trawling as a graduate student working on a paper. I was shocked to find that it was widely discussed by marine ecologists, but not so much by maritime archaeologists. The topic stayed in my mind long after writing the paper and finishing the class. Once I began working in underwater cultural heritage, I came back to the issue and decided, as many researchers do, to bring it to a conference. And so, thank you to Claire Zak and Sheri Kapahnke both for hosting the session and the dynamic discussion on coastal UCH sites at the 2022 EAA meeting that sparked the idea to pursue this research as well as for all the productive conversations that followed.

As editor for this volume and project leader for the 'Threats to Our Ocean Heritage Project', I need to highlight the incredible amount of help and assistance that has gone into this book. It would not have been possible without the research and expertise of so many individuals and organisations in the fields of maritime archaeology, cultural heritage, natural heritage, heritage law, and more.

Thank you to everyone at The Ocean Foundation who supported and made this project possible, from the first ideas to the UN Decade for Ocean Science endorsement, and all the help in between. The Ocean Foundation is the only community foundation for the ocean dedicated to reversing the trend of destruction of ocean environments around the world, and it is one of the few ocean NGOs that focuses on both natural and cultural heritage, making them uniquely situated to bring this project to light. Thank you to the president, Mark Spalding, for investing in underwater cultural heritage and the research that has made this possible. A particular thank you to the rest of the team at TOF who have helped us publicise, share, and organise everything to do with this project. Thank you to Bobbi-Jo Dobush and Madeline Warner for their incredible work fighting for a moratorium on Deep Seabed Mining, teaching me about the deep sea ecosystem, and helping fight to include UCH in the discussion. And finally, thank you to Ole Varmer, Senior Advisor on Ocean Heritage, for believing our project was worthwhile and the immeasurable help to get to this point.

This book, and the entire 'Threats to Our Ocean Heritage' project, would not be possible without the generous funding and assistance from the Lloyd's Register Foundation Heritage and Education Centre. Thank you to their director, Alex Stitt, for supporting the project and helping design it in the crucial early stages and to the rest of the team at LRF for their help and support. Without the help of Louise Sanger, Head of Research, Interpretation, and Engagement, and Benjamin Ferrari this book, nay this entire project, would not be what it is today.

I also want to thank all the other partners, researchers, idea-makers, and general assistance givers that have helped with this volume. I'm thrilled that the project is an endorsed Action under the UN Decade of Ocean Science for Sustainable Development and part of the amazing Ocean Decade Heritage Network and Cultural Heritage Framework Platform. Thank you to Athena Trakadas, chair of the ODHN, and Georgia Holly, project manager for CHFP, for their help with the endorsement, book, and our participation at global conferences. We have also been fortunate to work with The International Committee on Underwater Cultural Heritage (ICUCH) within the International Committee on Monuments and Sites (ICOMOS) on the project and are very grateful for their support and guidance.

A massive thank you to Michael Brennan, the wonderful editor of *Threats to Our Ocean Heritage: Potentially Polluting Wrecks*, contributor to this volume, and an all-around inspirational help to archaeologists and heritage professionals concerned about bottom trawling. Thank you to the reviewers of this book, whose insights and suggestions have made it a stronger contribution to the field and to the editing team at Springer for helping us bring this issue to light. And finally, thank you to my friends, family, and colleagues for the support along the way.

Charlotte Jarvis

Contents

1. **Introduction**... 1
   Charlotte Jarvis

2. **History of Trawling and Ecological Impact**...................... 9
   Charlotte Jarvis and Michael L. Brennan

3. **The Duty to Protect Our Ocean Heritage from Bottom Trawling**... 27
   Maria Pena Ermida

4. **Quantification of Bottom Trawl Damage to Ancient Shipwrecks: A Case Study from the Coastal Waters of Turkey**..... 41
   Michael L. Brennan

5. **Wreck Sites as Systems Disrupted by Trawling**.................. 53
   Jan Majcher, Rory Quinn, Gert Normann Andersen, and David Gregory

6. **Multi-use and Multi-vocal Challenges of Preserving Underwater Cultural Heritage in Stellwagen Bank National Marine Sanctuary**...................................... 67
   Calvin Mires, Benjamin Haskell, and Kirstin Meyer-Kaiser

7. **Underwater Cultural Heritage and Fishing Communities: Safeguarding Heritage and Safeguarding Fishers**............... 83
   Benjamin Ferrari, Antony Firth, David Gregory, and Louise Sanger

8. **Beneath a Façade: The Unscientific Justification of Treasure Salvage**.. 97
   Michael L. Brennan

9. **Conclusion: Looking Forward**................................... 107
   Charlotte Jarvis

Abbreviations

ABMT	Area-Based Management Tools
ABNJ	Areas Beyond National Jurisdiction
AIS	Automatic Identification System
BBNJ	Agreement Under the United Nations Convention on the Law of the Sea on the Conservation and Sustainable Use of Marine Biological Diversity of Areas Beyond National Jurisdiction
CFP	Common Fisheries Policy (EU)
CS	Continental Shelf
Defra	Department for Environment, Food and Rural Affairs (UK)
EU	European Union
EEZ	Exclusive Economic Zone
FMP	Fisheries Management Plan (UK)
FAO	Food and Agriculture Organization of the UN
HE	Historic England
IUU	Illegal, Unreported, and Unregulated Fishing
IFCA	Inshore Fisheries and Conservation Authorities
ITLOS	International Tribunal for the Law of the Sea
IUCN	International Union for the Conservation of Nature
LOSC	Law of the Sea Convention
LRF	Lloyd's Register Foundation
MMO	Marine Management Organisation (UK)
MPA	Marine Protected Area
MUCH	Maritime and Underwater Cultural Heritage
MHE	Maritime Heritage Ecology
NM	Nautical Miles
NMSS	National Marine Sanctuary System (US)
NOAA	National Oceanic and Atmospheric Administration
NRHP	National Register of Historic Places (US)
ONMS	Office of National Marine Sanctuaries (US)
RFMO	Regional Fisheries Management Organisation

SBNMS	Stellwagen Bank National Marine Sanctuary
SDG	Sustainable Development Goal
UCH	Underwater Cultural Heritage
KMA	UK Marine Area
UN	United Nations
UNGA	United Nations General Assembly
VME	Vulnerable Marine Ecosystem

List of Figures

Fig. 1	The wreck of a barge, likely damaged by bottom trawling...............	vii
Fig. 1.1	The relative intensity of all types of fishing around the coast	3
Fig. 2.1	A beam trawl (top) and otter trawl (bottom)....................................	10
Fig. 2.2	Rockhopper gear (left) and position on a trawling net (right)	11
Fig. 2.3	Trawling net with tickler chains towed ahead of net	12
Fig. 2.4	Scallop dredge (left) and row of scallop dredges (right)	13
Fig. 2.5	An example of shipwrecks off the coast of Africa............................	17
Fig. 3.1	The maritime zones include the 12 nm Territorial Sea, the 24 nm Contiguous Zone, the 200 nm Exclusive Economic Zone, and the High Seas	28
Fig. 4.1	ROV image of Marmaris B shipwreck..	45
Fig. 4.2	Graph showing extent of trawl damage by distance from shore in terms of % broken amphoras......................................	47
Fig. 4.3	Two sonar surveys of Eregli E	48
Fig. 5.1	Conceptual diagram for wreck site dynamics for (**a**) high-energy and (**b**) low energy environments, along with (**c**) corresponding estimates for site preservation	55
Fig. 5.2	Multibeam echosounder data-derived, hillshaded digital elevation models of (**a**) a steamship wreck, possibly *SS Edgar* in the Baltic Sea with (**b**) a bathymetric profile denoted as AA'. (**c**) hillshaded digital elevation model of an unknown wreck in the Danish North Sea	57
Fig. 5.3	(**a**) Differences between multibeam echosounder point clouds obtained for the SS *WM Barkley* site in 2015 and 2019. (**b**) CFD-simulated NNW-oriented tidal current flow pressure exerted on the modelled seabed at the same site, (**c**) vorticial patterns delineated by the simulations, (**d**) NNW-oriented current flow streamlines, and (**e**) wall shear stress	60

Fig. 5.4	(**a**) Digital Elevation Model of Differences (DoD) for the FV *St. Michan* site, showing bathymetric changes between years 2015 and 2019, and (**b**) inset map showing a hillshaded elevation model created using the data collected in 2019	62
Fig. 6.1	Map of Stellwagen Bank National Marine Sanctuary	68
Fig. 6.2	Heatmap with shipwrecks overlayed on areas of fishing activity	69
Fig. 6.3	The fishing vessel, *North Star*, in 2016 before (left) and after a scallop derby in 2017 (right)	73
Fig. 6.4	Top: Sidescan of *Portland* in 2010 with intact stern; Middle: In 2021 with stern missing; Bottom: Photogrammetric model of *Portland* showing fishing gear draped over and around damaged stern area	74
Fig. 6.5	Side-scan image of trawl marks through the possible *King Philip* shipwreck site (left); and through the ship's starboard bow section (right)	75
Fig. 6.6	Biological communities and impacts of fishing gear entanglement	75
Fig. 6.7	Biological communities and impacts of fishing gear entanglement on *Portland*	76
Fig. 6.8	Outreach bulletin requesting voluntarily avoidance of listed wreck sites (left); placards that could be placed in wheelhouses as avoidance reminders (middle and right); a QR Code (middle of right-hand panel) takes fishers to a NOAA Fisheries Web site where they can see the site locations and download coordinates to upload into their chart plotters	77
Fig. 7.1	Fishing intensities. (**a**) Bottom beam trawling, (**b**) bottom otter trawling, (**c**) bottom seines, (**d**) total bottom fishing intensity	86

List of Table

Table 6.1　List of Identified Shipwrecks within Stellwagen Bank National Marine Sanctuary 71

Chapter 1
Introduction

Charlotte Jarvis

Trawling has been recognised as a profoundly damaging practice with lasting negative consequences on seabed ecology and marine life since its first mention in a 1376 parliamentary petition. Mobile fishing gear (including any dredge, trawl, or similar device) is used to tow or push a net with a boat to catch fish. Bottom trawling, in particular, grew from a need to keep up with declining fish stocks and developed further with technological changes and increased demands, though it created 'anger and resentment' within the fishing communities (Bolster, 2012, p. 236). All three trawling revolutions—invention, mechanisation, and later deep-water expansion—have been met with controversy and pushback by the public and environmentalists alike (Roberts, 2008).

There is no doubt that trawling has decimated fish stocks globally which has brought hardship on fishing communities. Today's boats must work an estimated 17 times harder than in the past because there are literally fewer fish in the sea (Roberts, 2012; Thurstan et al., 2010). Every year trawlers plough areas of the seabed roughly equal to half of the world's continental shelves and convert the rich seafloor below into a bleak landscape of flat nothing (Watling & Norse, 1998). With one pass by a trawler, boulders can be displaced, large epifaunal invertebrates removed and damaged, and sediment re-suspended (Freese et al., 1999). Yet little has been done to seriously limit the practice and protect the underwater landscape.

Crucially, archaeological impacts and data are also missing from biological reports and published articles on the practice. Trawling also has dramatic impacts on maritime archaeology sites, though these effects are less well known. The seafloor landscape includes historical information and sites of cultural significance. Bottom trawling does not just destroy the physical fish habitats—important shipwrecks and artefacts are lost too and ostensibly have been since the inception of trawling. Trawling gear damages wrecks when their nets snag and are tangled with

---

C. Jarvis (✉)
The Ocean Foundation, Washington, DC, USA

shipwreck structures, often destroying components, and the trawl doors and chains can destroy and scatter artefacts and vessel components (Brennan, 2016). Archaeologists have recently begun raising awareness about the impact of trawling on their sites, and more work is needed.

Although legislation that limits trawling can help biological communities rebound, the archaeological material lost can never be recovered. And although efforts to illustrate sustainability in seafood, such as the Marine Stewardship Council stamp, may attempt to show safe fishing practices, one can never be sure of the wider harm being done, especially to Underwater Cultural Heritage (UCH). Despite decades of research and protests against trawling from fishery scientists and marine ecologists, archaeologists have only recently begun examining this industry's impacts to UCH sites.

This book serves as a call for action to address the threats to our Ocean Heritage from the destructive activities of bottom trawling. In the follow up to his seminal work *The Unnatural History of the Sea*, conservationist Callum Roberts writes in *The Ocean of Life* that the 'impacts [of trawling in this case] are discussed in isolation at different meetings and by different people, who never quite see the overall picture' (Roberts, 2012, p. 6). He is referring to different people within the marine scientific fields such as those focused on fishery sciences or pollution. His point holds true, though, with a group of people many do not consider: Ocean Heritage professionals. Underwater Cultural Heritage is an integral part of both cultural heritage and natural heritage, thus making it a shared **Ocean Heritage** which must be considered part of a Marine Spatial Planning programme that should ban bottom trawling at UCH sites.

1.1 The Importance of Our Ocean Heritage

1.1.1 Cultural Heritage

The stories of our societies and our ancestors are wrapped in intangible connection with the ocean and preserved on the seafloor as artefacts, shipwrecks, and remains of those lost or buried at sea. Marine global heritage has largely been based on natural features, and the inextricable link between natural and cultural has been ignored. However, UCH functions in a shared space of natural and cultural heritage: Ocean Heritage. Bottom trawling impacts this Ocean Heritage and harms both natural and cultural objects on the seafloor.

Coastal trawling is particularly destructive to UCH since it is where most known wrecks are located and is where nearly all submerged landscapes, the continental shelf where submerged prehistoric sites are located from time periods of lower water levels, are situated (Evans et al., 2014; Bailey et al., 2017). However, there is also a massive amount of ecological damage from trawling in coastal waters. Coastal seas are some of the most ecologically productive areas, with delicate plants, marine life, and complex ecological systems. A 2021 report found that trawling is most

1 Introduction

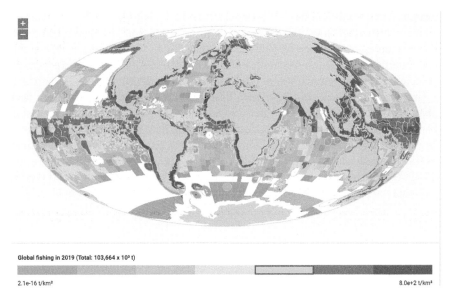

Fig. 1.1 The relative intensity of all types of fishing around the coast is illustrated here in tonnes per square kilometre. (Source: Pauly D., Zeller D., Palomares M.L.D. (Editors), 2020. Sea Around Us Concepts, Design and Data (seaaroundus.org) printed under Creative Commons Attribution Non-Commercial 4.0 International License)

intense in the territorial (12 nautical miles offshore) seas of coastal states (Steadman et al., 2021; Pauly et al., 2020) (Fig. 1.1).

As technology improves, archaeological investigation will move further out to the deep sea, and those sites must be protected from trawling too. The deep-water sites are also valuable resources since, being inaccessible for so long, they have had the least anthropogenic damage. Remotely Operated Vehicle (ROV) technology has allowed previously impossible deep-sea exploration and has revealed both beautiful, complex habitats as well as shipwrecks that have been already damaged by trawlers. For example, in the Black Sea and Turkey's Mediterranean, Aegean, and Black Sea shores (Brennan et al., 2016), there is evidence of trawl damage on deep-water wrecks. Ereğli E, one of the oldest deep-water wrecks yet discovered in the Black Sea, could have been one of the most significant wrecks in the area, but by the time it was found in 2011, it had already been nearly destroyed by trawling (Brennan et al., 2016).

1.1.2 Natural Heritage

Trawling also does extensive damage to the seafloor and ecosystems within it, as will be discussed in Chap. 2. The effects have been likened to clear cutting a forest. Structural components of fish habitats are removed, and biodiversity is lost. For example, the North Sea has been heavily fished for centuries, and this increased significantly in the 1900s (Frid et al., 2000). Information from published and unpublished sources dating back 60 years shows that fishing has influenced the benthic communities of the North Sea, with a definite change in composition of the benthos coinciding with increased rates of trawling (Frid et al., 2000).

Important deep-sea sites have been destroyed before scientists have even been able to study them. In deeper waters, fauna is less adapted to changes and disturbances to their environment. The effects of trawling here take decades, not years, to be recovered (Jones, 1992). The morphology of the seafloor is changed over time, as trawl gear displaces sediments and reduces the original complexity (Puig et al., 2012). Bottom trawling has become a driver of deep seascape evolution and is comparable to areas destroyed by large-scale agricultural ploughing (Puig et al., 2012).

1.1.3 Underwater Cultural Heritage as Both Natural and Cultural

UCH can support ecological marine biodiversity and helps boost sea connectivity. A paper by Kirstin S. Meyer-Kaiser and Calvin H. Mires coined the term 'Maritime Heritage Ecology', a field that must be further explored (Meyer-Kaiser & Mires, 2022) and the importance of which is highlighted in chapter 6. This is because cultural heritage and natural heritage are intertwined. UCH can support ecological marine biodiversity and helps boost sea connectivity. Natali Pearson and Benjamin Thompson argue that it is beneficial for sites with high UCH and high natural heritage to co-occur and be used strategically together (Pearson & Thompson, 2023).

Over time, shipwrecks come into equilibrium with the marine environment around them and become part of the benthic topography (Brennan et al., 2016). Trawling destroys this equilibrium, which often has been balanced for centuries (Brennan et al., 2013). Shipwrecks function as artificial reefs and are crucial to habitats. They provide shelter for juvenile species, add hard materials to an otherwise soft seafloor, and can be as indispensable to the seafloor ecology as a natural coral reef or seamount. Waters around protected wrecks have been documented to produce higher fish population and biomass than in areas of high trawling with no structures on the seafloor (Brennan, 2016; Krumholz & Brennan, 2015). Through this process of 'spill over', protected shipwrecks can help increase the strength of surrounding fish stocks. Thus, shipwrecks should not be viewed solely in a cultural significance context. They are part of the natural ocean landscape as well and our Ocean Heritage.

Trawling can also cause damage to potentially polluting shipwrecks (PPWs) and further hurt the seabed ecosystem (Brennan et al., 2023). The topic of PPWs is covered in Michael Brennan's edited volume *Threats to Our Ocean Heritage: Potentially Polluting Wrecks* and a recent article highlight, through case studies, the risk to oil-containing shipwrecks posed by trawling (Brennan et al., 2023).

1 Introduction

It is difficult to map the history of trawling on sites, as many visible trawl scars on the seabed fade over time, rendering them invisible to an archaeologist investigating the site. There have been a few regional studies in recent years, though, that clearly illustrate the danger faced by historic sites. In this book, after an introduction to the history of bottom trawling and the ecological damage caused by mobile fishing gear, Maria Peña Ermida highlights the legal context within which bottom trawling sits, and advocates for the precautionary principle[1] and duty to protect UCH within the marine environment.

Next, case studies from some archaeological sites destroyed by trawling are used to illustrate the destructive nature of the practice. Michael Brennan begins the case study section with an overview of his extensive work highlighting trawl damage in the Mediterranean and discussing more recent scholarship. The work done by Brennan and the team in Turkey presents a method 'to truly quantify change to a shipwreck site over time by bottom trawling' by conducting return visits and repeated surveys to sites. It is proposed that 'more work like this is needed to illustrate the threat posed to shipwrecks'.

The chapter from Jan Majcher, Rory Quinn, Gert Normann Andersen, and David Gregory explores the impact of bottom trawling on individual shipwreck sites. The authors discuss shipwrecks as process-response systems, which over time reach a state of quasi-equilibrium that can be disrupted by the physical disturbance of bottom-contact fishing. They draw on examples from recent geophysical surveys carried out in the Irish, Baltic, and North Seas, examining the potential of contemporary geophysical techniques in tracking changes at underwater sites induced by bottom trawling. Furthermore, knowledge gaps related to the impact of bottom-contact fishing on UCH are identified, as well as the challenges associated with addressing them. It can be difficult to directly detect evidence of disturbance on dynamic seafloors, but detailed examination of wreck distribution may provide some clues to whether structural damage to the ship can be attributed to bottom trawling. The chapter concludes with the authors advocating for more case studies to better understand the impact of fishing on shipwreck sites.

Calvin Mires, Benjamin Haskell, and Kirstin Meyer-Kaiser use a US National Marine Sanctuary, Stellwagen Bank (SBMNS), as a case study to show the complexities of protecting UCH at risk for trawling while also working within unique situational requirements. SBMNS is in New England, where fishing activities are an integral part of many people's livelihoods and have been since long before the sanctuary. Additionally, as a national sanctuary, Stellwagen's management is also bound to facilitate public and private activities, like fishing, while also protecting the sanctuary's resources, like Ocean Heritage. Along with providing images that show trawling's negative impacts on both natural and cultural heritage, the authors also deliver a case study that highlights the voice of fishers in the area, and the needs of all stakeholders. They show the necessity to integrate maritime historical ecology into management systems because the Ocean Heritage that needs protection

---

[1] Also often cited as the precautionary approach.

encompasses both natural and cultural resources. They also suggest that 'UCH offers many opportunities to understand ocean processes, and its archaeological and ecological roles are intertwined'. Maritime Heritage Ecology (MHE) is part of a solution for Stellwagen's trawling management dilemma.

Ben Ferrari, Antony Firth, David Gregory, and Louise Sanger also discuss a road to safeguard Ocean Heritage from trawling in a way that, crucially, also safeguards fishers. They highlight the 'shared concerns between heritage managers and other stakeholders in the context of a complex marine management and stewardship environment', and the needs for both evidence-based characterisations of the impacts of fishing, and their integration into consideration of site formation. They introduce three exciting new initiatives focused on this: The European Research Council's Project ENDURE, Lloyd's Register Foundation's work on UCH protection in the context of engineering a safe, sustainable ocean economy, and Historic England's work to engage with fishers to mitigate damage to UCH and develop better management regimes. They tie their work to the notion of a rapidly growing Blue Economy and the future strains that will be placed on ocean systems in the decades to come. This leads to the conclusion of this volume, where possible steps forward are discussed and a call to action is made.

1.2 Steps Forward

Underwater policies formulated to manage offshore fishing based on cultural preservation are lacking. Archaeologists are significantly behind benthic ecologists and fishery management professionals in advocating for regulation. Some trawling restrictions have been placed after backlash in the 1990s when ecologists and fisheries experts, well aware of the dangers of trawling, lobbied for more restrictions. This research and advocacy for regulation is a good start but needs more activism led by archaeologists. There is a preferred policy for *in situ* preservation considered as a first option in the 2001 UNESCO Convention but some practical measures for site managers to address the threats from bottom trawling are needed. If *in situ* preservation is to be supported, moorings should be added so that shipwrecks, when left in place, can become artificial reefs. However, what is needed most is for states and international fishing organisations to ban bottom trawling at and around identified UCH sites, as has already been done for some seamounts. Most importantly, trawling damage should not be allowed as a justification for salvage by treasure-hunting companies either, as Michael Brennan discusses in Chap. 8.

These results cannot be achieved without the voice and support of the fishing community. Wrecks can also present problems to fishers. When nets snag, gear is lost or ruined and can be costly to replace. Shipwrecks can also endanger the lives of those onboard vessels. In Britain, from 1976 to 2005, around 16 vessels capsized after gear was snagged on obstructions, and many lives were lost (Atkinson, 2012). Although some shipwreck sites have only been found because of fishers' reports, there is no requirement for them to do so. A clearer partnership between archaeologists and fishers is needed and archaeologists must understand the context fishers operate in to communicate effectively.

The steps we take to minimise trawling damage to Ocean Heritage can pave the way for other important ocean exploitation. Climate change will continue to threaten our ocean (for example, sea-level-rise sinking previously terrestrial sites) and we know ecologically why it is important to protect the ocean. While there are many unknowns regarding deep-sea biodiversity and ecosystem services, what we do know clearly points to vast and far-reaching damage. In other words, we know enough from the existing trawling damage that tells us we should stop similar practices, like seabed mining, immediately.

We must use precautionary measures, explained by Maria Peña Ermida in Chap. 3 and not start further exploitative practices. This is particularly important when talking about the deep-sea, since it is often left out of conversations about the ocean, which in turn have been left out of discussions about climate and environment. In fact, these things are all crucial and deeply connected. We cannot predict what sites are yet to be discovered and which are historically significant. Trawling is a danger, both to fish populations and habitats, as well as cultural landscapes. There should not be a choice between humans and the natural world. Restrictions must be passed as soon as possible to prevent further destruction to our Ocean Heritage.

References

Atkinson, C. M. (2012). *Impacts of bottom trawling on underwater cultural heritage* [Master's thesis]. Texas A&M University.
Bailey, G., Harff, J., & Sakellariou, D. (2017). *Under the sea: Archaeology and palaeolandscapes*. Springer Berlin Heidelberg.
Brennan, M. (2016). Quantifying impacts of trawling to shipwrecks. In M. E. Keith (Ed.), *Site formation processes of submerged shipwrecks* (pp. 157–179). University Press of Florida.
Brennan, M. L., Davis, D., Roman, C., Buynevich, I., Catsambis, A., Kofahl, M., Ürkmez, D., Ian Vaughn, J., Merrigan, M., & Duman, M. (2013). Ocean dynamics and anthropogenic impacts along the southern Black Sea shelf examined through the preservation of pre-modern shipwrecks. *Continental Shelf Research, 53*, 89–101. https://doi.org/10.1016/j.csr.2012.12.010
Brennan, M. L., Davis, D., Ballard, R. D., Trembanis, A. C., Vaughn, J. I., Krumholz, J. S., Delgado, J. P., Roman, C. N., Smart, C., Bell, K. L. C., Duman, M., & DuVal, C. (2016). Quantification of bottom trawl fishing damage to ancient shipwreck sites. *Marine Geology, 371*, 82–88. https://doi.org/10.1016/j.margeo.2015.11.001
Brennan, M. L., Delgado, J. P., Jozsef, A., Marx, D. E., & Bierwagen, M. (2023). *Site formation processes and pollution risk mitigation of World War II oil tanker shipwrecks: Coimbra and Munger T. Ball* [Preprint]. In Review. https://doi.org/10.21203/rs.3.rs-3029829/v1
Bolster, W. J. (2012). *The mortal sea: Fishing the Atlantic in the age of sail*. Belknap Press of Harvard Univ. Press.
Evans, A. M., Flatman, J., & Flemming, N. C. (Eds.). (2014). *Prehistoric archaeology on the continental shelf: A global review*. Springer.
Freese, L., Auster, P. J., Heifetz, J., & Wing, B. L. (1999). Effects of trawling on seafloor habitat and associated invertebrate taxa in the Gulf of Alaska. *Marine Ecology Progress Series, 182*, 119–126. https://www.jstor.org/stable/24852124

Frid, C. L. J., Harwood, K. G., Hall, S. J., & Hall, J. A. (2000). Long-term changes in the benthic communities on North Sea fishing grounds. *ICES Journal of Marine Science, 57*(5), 1303–1309. https://doi.org/10.1006/jmsc.2000.0900

Jones, J. B. (1992). Environmental impact of trawling on the seabed: A review. *New Zealand Journal of Marine and Freshwater Research, 26*(1), 59–67. https://doi.org/10.1080/00288330.1992.9516500

Krumholz, J. S., & Brennan, M. L. (2015). Fishing for common ground: Investigations of the impact of trawling on ancient shipwreck sites uncovers a potential for management synergy. *Marine Policy, 61*, 127–133. https://doi.org/10.1016/j.marpol.2015.07.009

Meyer-Kaiser, K. S., & Mires, C. H. (2022). Underwater cultural heritage is integral to marine ecosystems. *Trends in Ecology & Evolution, 37*(10), 815–818. https://doi.org/10.1016/j.tree.2022.06.014

Pauly, D., Zeller, D., & Palomares, D. (2020). *Sea around us concepts, design and data*. seaaroundus.org

Pearson, N., & Thompson, B. S. (2023). Saving two fish with one wreck: Maximizing synergies in marine biodiversity conservation and underwater cultural heritage protection. *Marine Policy, 152*, 105613. https://doi.org/10.1016/j.marpol.2023.105613

Puig, P., Canals, M., Company, J. B., Martín, J., Amblas, D., Lastras, G., Palanques, A., & Calafat, A. M. (2012). Ploughing the deep sea floor. *Nature, 489*(7415), 286–289. https://doi.org/10.1038/nature11410

Roberts, C. (2008). *The unnatural history of the sea* (1st paperback ed.). Island Press/Shearwater Books.

Roberts, C. (2012). *The ocean of life: The fate of man and the sea*. Viking.

Steadman, D., Thomas, J., Villanueva, V., Lewis, F., Pauly, D., Palomares, M. L., Bailly, N., Levine, M., Virdin, J., Rocliffe, S., & Collinson, T. (2021). *New perspectives on an old fishing practice: Scale, context and impacts of bottom trawling*. https://doi.org/10.13140/RG.2.2.12318.41286/1

Thurstan, R. H., Brockington, S., & Roberts, C. M. (2010). The effects of 118 years of industrial fishing on UK bottom trawl fisheries. *Nature Communications, 1*(1), 15. https://doi.org/10.1038/ncomms1013

Watling, L., & Norse, E. A. (1998). Disturbance of the seabed by mobile fishing gear: A comparison to forest clearcutting. *Conservation Biology, 12*(6), 1180–1197. https://www.jstor.org/stable/2989836

Open Access This chapter is licensed under the terms of the Creative Commons Attribution 4.0 International License (http://creativecommons.org/licenses/by/4.0/), which permits use, sharing, adaptation, distribution and reproduction in any medium or format, as long as you give appropriate credit to the original author(s) and the source, provide a link to the Creative Commons license and indicate if changes were made.

The images or other third party material in this chapter are included in the chapter's Creative Commons license, unless indicated otherwise in a credit line to the material. If material is not included in the chapter's Creative Commons license and your intended use is not permitted by statutory regulation or exceeds the permitted use, you will need to obtain permission directly from the copyright holder.

Chapter 2
History of Trawling and Ecological Impact

Charlotte Jarvis and Michael L. Brennan

2.1 The History of Trawling

Bottom trawling developed from the practice of fishing with a net or long lines, but that older style was able to be more controlled and discriminate in its catch. With bottom trawling, nets are weighted to keep them low along the seafloor and a large beam on deck spreads nets wide to increase catch. Marine biologists and environmentalists have been concerned about trawling since it began. As far back as 1376, the English parliament highlighted the destructive nature of the practice to fish populations and habitats, noting that it 'runs so heavily and hardly over the ground when fishing that it destroys the flowers of the land' and takes so many fish 'to the great damage of the commons and the destruction of the fisheries' (Petition by the Commons to King Edward III, 1376 seen in Bolster, 2012). This first known mention of trawling calls for the removal of the practice and yet the activity endures. Trawling in Europe continued sporadically through the medieval and early modern periods using mainly the beam trawl method until the early nineteenth century when the industrial revolution pushed the technology further.

By 1840, fishers using sail power were already noticing depleted stocks, so changes developed out of a necessity to compensate for a declining catch. The introduction of steam-powered vessels ushered in a new modern era of commercial fishing (Roberts, 2012). The first purpose-built British steam trawling vessels were completed in 1881 and could gather four times more fish per catch than a sailing vessel (Bolster, 2012). Steam powered vessels were not as limited by weather and

C. Jarvis (✉)
The Ocean Foundation, Washington, DC, USA

M. L. Brennan
SEARCH Inc., Jacksonville, FL, USA
e-mail: mike@brennanexploration.com

© The Author(s) 2024
C. Jarvis (ed.), *Threats to Our Ocean Heritage: Bottom Trawling*,
SpringerBriefs in Underwater Archaeology, https://doi.org/10.1007/978-3-031-57953-0_2

had the strength to power nets with chain reinforcements on the seafloor. By the end of the nineteenth century, this enabled the transition to the otter trawl, where two 'doors' are used to open the nets and accommodating larger nets (see Fig. 2.1 for a comparison of the methods) (National Research Council, NRC, 2002). Net size was no longer limited by the length of the vessel or the wooden beam, but by the power of the tow vessel (Atkinson, 2012). Beam trawling remained popular in some European countries, though and while otter trawl use increased, beam trawls were never fully abandoned (Ferrari, 1995).

In the United States, scientists initially welcomed trawling as it gathered more fish to study and, since they believed the sea to be limitless, they were unconcerned about the practice (Bolster, 2012). Beam trawlers were introduced just before the twentieth century in Cape Cod and soon gasoline powered engines revolutionised the industry (NRC, 2002). In 1912, Congress appointed funds to the Bureau of Fisheries to investigate 'whether or not this method of fishing is destructive to the

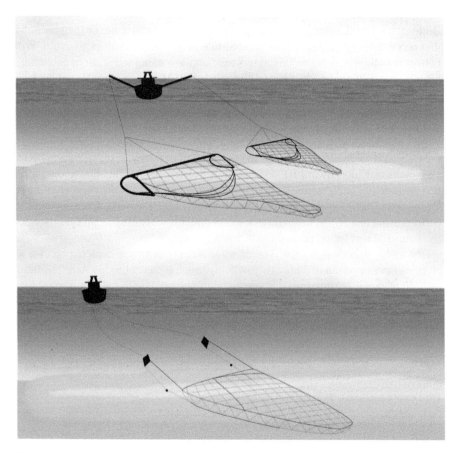

Fig. 2.1 A beam trawl (top) and otter trawl (bottom). (Source: Ecomare/Oscar Bos, licensed under the Creative Commons Attribution-Share Alike 4.0 International)

fish species or is otherwise harmful' (Bureau of Fisheries, 1914 and quoted in Bolster, 2012, p. 258). Two and a half years later, the report was published and, in an effort to prevent monopolies, did not recommend the prohibition of the practice and opened the door to large-scale otter trawling in 'certain definite banks and grounds', though it admitted 'While the facts before us show no proof or presumption of any depletion of the fisheries on the banks frequented by American otter trawlers… it is possible that the seeds of damage already have been sown and that their fruits may appear in the future' (Bureau of Fisheries, 1914 and quoted in Bolster, 2012, p. 258).

Arguably the most detrimental technological advance came after World War II. Factory-sized, freezer-equipped trawlers were introduced that had stern ramps to accommodate larger nets, quick freeze facilities and storage, and fishmeal processing machines onboard. The increase in cod catches off Newfoundland are a good example to illustrate the effects. From 1875 to 1955, steam and gasoline power caused catches to rise from 160,000 to 300,000 metric tons. New factory trawlers brought in 500,000 tons in 1960 and 800,000 tons in 1967. These examples also illustrate how quickly stocks can become depleted, however, because over the next 9 years, catches fell to 150,000 tons (Bolster, 2012).

Trawling spread to greater depths, spearheaded by the Soviet Union, and it quickly grew into a global industry. Previously, the high seas were regarded as too dangerous and not economically worth fishing, but the 1950s and 1960s saw the rise of echosounders and increased access to areas beyond the reach of shallow trawls. But these new fishing grounds often had rough seafloors that shredded the nets. In response, a technology was developed to shield the nets from obstructions, but which also significantly increased the destructive power of trawls. 'Rockhopper gears', measuring up to 1 m in diameter and weighing hundreds of kilograms, are massive steel rubber rollers attached to ground ropes on trawl nets, enabling the nets to roll over obstacles, but which increase the damage to the seabed (He et al., 2021; Watling & Norse, 1998). 'Tickler chains' would also be added ahead of the ground rope to scrape the seafloor and drum up catch (Jones, 1992; Watling & Norse, 1998) (Figs. 2.2 and 2.3).

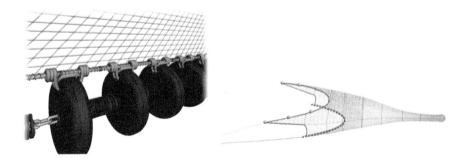

Fig. 2.2 Rockhopper gear (left) and position on a trawling net (right). (Source Seafish: www.seafish.org/ reprinted with permission)

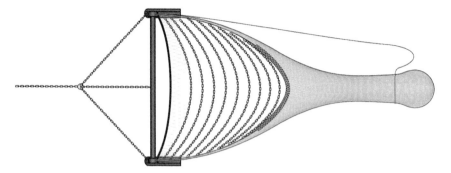

Fig. 2.3 Trawling net with tickler chains towed ahead of net. (Source Seafish: www.seafish.org/ reprinted with permission)

Commercial fishers now had the technology to send their boats out for longer, fish further away, and fish faster with larger nets (NRC, 2002). Eastern European nations took these to a new level creating 'floating towns built for the sole purpose of processing marine life into food' (Roberts, 2008, p. 189). Biologically productive deep seamounts became the new targets and using sonar, vessels could trawl through a large group of fish taking in 50–60 tonnes in a few minutes (Roberts, 2008).

Today, trawling has grown to previously unimaginable rates. The nets used, some measuring more than 4 stories tall, cut paths through the seafloor at least 100 m wide, pulling in fish at unsustainable rates—a recent FAO report shows that one-third of global fish stocks are overfished (FAO, 2020). A study by Villy Christensen and his team estimated that today's fish stocks are a tenth of what they were in 1900, and two thirds of that destruction comes from 1950 onwards (Christensen et al., 2003). But, as Callum Roberts shows, by 1900 stocks were already heavily decreased so the real number is more likely less than 5% of natural levels (Roberts, 2008).

Many nets are made of synthetic materials which do not break easily when snagged, allowing for higher powered vessels (Brennan, 2016). These nets also make up the largest portion of marine debris and plastic pollution in the ocean (Napper et al., 2022). With modern improved gear and electronics, vessels use sonar to target schools of fish and specific bathymetric areas, allowing for fishing over targeted areas of seabed, and drag their nets over rougher seafloor than before, decimating fish populations to a greater extent (Pederson & Dorsey, 1997).

2.1.1 Dredging

Trawling is not the only method of bottom fishing, and in many cases the impacts of dredging on the seabed are just as harmful as bottom trawling. Dredging and trawling are grouped together as fishing methods known as 'mobile fishing gear' and though bottom trawling is the focus of this book, dredging will be briefly summarised. A dredge is a cage-like contraption, sometimes with a scraper blade or teeth on the lower part, which is towed behind the vessel to excavate organisms out of substrate and capture them (He et al., 2021). Molluscs, particularly mussels, oysters, scallops, and clams, are the most targeted species. Towed dredges or mechanised (hydraulic) dredges are used, and these methods have similar environmental impacts to trawling with both long- and short-term results including the elimination of natural bottom features and flattening of substrate, shift in surface sediment, burying of organisms, reduction in seagrass, and reduction in species abundance (Caddy, 1973; Currie & Parry, 1996; Hall et al., 1990; Hall-Spencer, 2000; Kaiser, 1997). Dredges also carve deeper furrows in the seabed than trawls, as their target species are within the sediment rather than on top of it (Fig. 2.4).

The historical origins of dredging are unknown but seem to predate trawling in some areas (Ferrari, 1995). The impact on UCH is similar to bottom trawling and many surveys and research mention dredging activity in the area (see Steinmetz's thesis examining trawling and dredging impacts on shipwrecks in the Mid-Atlantic 2010). For example, off the Isle of Mull, Scotland, the wreck of the Swedish general-cargo steamship SS *Hispania* (built 1912) has been salvaged and visited by divers since the 1950s. In 1999, a scallop dredger caught on the vessel and the scrapes made by the gear damaged the hull as well as laying a mast on its side (Robertson, 2007). Additional surveys in 2001 showed further dredge damage. Dredging also poses a significant risk to UCH found in the waters of the Arctic (Ragnarsson et al., 2006). This is especially worrying to archaeologists because of the high level of preservation exhibited by shipwrecks in this location (Nœvstad, 2006). Like in the deep-sea, shipwrecks that have been undiscovered and well-preserved are damaged by fishing activities before they can be studied. For example, scallop dredgers near Spitsbergen have pulled anchors, ship timbers, and ceramics from the ocean and with only a few incidents actually reported, it is difficult to know the full extent of damage (Nœvstad, 2006). Dredge impacts have also been documented on steel hulled wrecks, such as that of the oil tanker *Coimbra*, sunk off Long Island in 1942 (Brennan et al., 2023).

Fig. 2.4 Scallop dredge (left) and row of scallop dredges (right). (Source Seafish: www.seafish. org/ reprinted with permission)

2.2 The Ecological Impact

Trawling has been shown to harm sea life and the sea floor by reducing topography, compression and resuspension of sediments, decreasing complexity, and causing both physical and chemical damage to the ecosystem. Trawling can penetrate sediment 20 cm or more and cut scars into the ground 1–3 m wide, causing advection and reoxygenation of the sediment strata (Friedlander et al., 1999). The otter trawls can rework the seafloor, sometimes moving boulders that weigh more than 30 tonnes (Atkinson, 2012). This decreases the roughness, and thus the complexity of the bottom, by smoothing the bottom structures and removing bottom fauna (Pederson & Dorsey, 1997; Brennan et al., 2012). Structures like ripples, mesophotic reefs, and other soft suberate habitats are destroyed (Brennan et al., 2016).

Coastal seas are some of the most ecologically productive areas and areas of heavy trawling. Marine ecologist Callum Roberts writes that 'the spread of trawling caused the greatest human transformation of marine habitats' (Roberts, 2008, p. 156). Seas have shifted from complicated, productive, rich environments to expanses of flat gravel and mud as continuous trawling in these fishing grounds has eliminated entire habitats. In the Wadden Sea, for example, reefs thousands of years old are gone and so are the bottom habitats, and fauna reliant on them (e.g., oyster reefs, eelgrass, seaweed) (Lotze, 2005).

Trawling also resuspends sediments, which is detrimental to the sedimentary and chemical processes occurring on the seafloor (Brennan et al., 2016). The nutrients and toxins within the sediment can alter the chemical composition of the habitat, which also can increase metal corrosion and organic material deterioration on shipwrecks (Duplisea et al., 2001; Mayer et al., 1991; Pilskaln et al., 1998). The resuspension and reoxygenation of the sediments disturb the anoxic bacterial processes occurring below the sediment-water interface and diminishes the benthic role of bacterial decomposition of organic matter within the surface sediments.

A more obvious impact of trawling is the decimation of fish populations. Intensive fishing alters the balance between the young and old of a population and when the balance is off, populations cannot recover. Areas that have been heavily trawled become dominated by small-bodied species that can colonise and recover quickly, displacing larger long-lived species (Olsgard et al., 2008). These smaller benthic species can withstand the mortality imposed by trawling and then benefit from the reduced competition in the ecosystem (Jennings et al., 2001).

The continuous intensive overfishing of target species makes it nearly impossible for populations to recover and reduces the ability for the fishery to be sustainable. This is shown by an example that during WWI and WWII, as the fishers and their boats were taken into the war services, some fish populations rebounded. In the North Sea, species benefited from the break in exploitation and when fishing resumed in peacetime, there were larger catches than previous (Beare et al., 2010; Holm, 2012). Inevitably, the populations crashed again, and fishers sought new grounds.

2 History of Trawling and Ecological Impact

The effects of trawling in deep water are more pronounced. Fish species in the deep ocean have lower growth rates and populations take longer to recover than in coastal waters. Early 1980s catches in the deep waters of the Gulf of Maine were twice as high as the levels that could be sustainable and by the mid-1980s, 60–80% of species were taken (Dobbs, 2000). Various estimates of the cod population in this area today put the numbers around one third of one percent of what it should be (Rose, 2004). Trawlers also commonly target deep seamounts, which function as 'refuelling stops' for many fish species in the open ocean. The populations here, and the benthic ecosystems, have been decimated by continuous overfishing.

Trawling has not necessarily helped fishers and the global fish market either. Intensive trawling, through seafloor destruction and population decimations, undermines food webs that support the fish species we seek to catch for consumption. Roberts estimates that fishers today pull in just 6% of what they did about a century ago, showing that there are literally fewer fish in the sea because of human practices (Roberts, 2012). A World Bank report, The Sunken Billions, noted that if we fished less, the major fish stocks globally would begin to produce 40% more in a few years and maybe 60% in European waters (Willmann & Kelleher, 2009). But fishing quotas set by EU policy leaders still tend to range one third higher than the safe catch levels recommended by their own scientists.

2.3 Regional Case Studies

Due to a number of factors, including accessibility and research conducted, not every region of the ocean could be covered as a case study in this volume. Some of the key findings and notable research are summarised below.

2.3.1 Pacific

The issue of heritage destruction by trawling is important to the Asia-Pacific Region because traditionally, trawling studies with UCH have focused on European and North American waters, even though roughly 50% (over the last decade of available data) of all bottom trawled fish come from the continental shelf/EEZs of Asian nations or the foreign fleets of Asian countries (Steadman et al., 2021, p. 5). Many wrecks in the Pacific region lie in areas of heavy trawling activity and the wreck timbers have already been damaged. For example, the Longquan Wreck is a fifteenth-century wreck with timbers standing nearly two metres tall in the deep water off the coast of Malaysia. It was originally found with a fairly intact cargo but was later found to have been flattened by Thai trawling vessels (Flecker, 2002). This was also the case with the sixteenth-century Singtai Wreck (Flecker, 2002).

Three kilometres west of Ko Si Chang, Thailand, the Chang I wreck was found in 1982 during the excavation of a nearby site (Green et al., 1986). The site consisted of scattered ceramic sherds and some exposed hull timbers. The wreck was visited in subsequent seasons and archaeological survey markers were left behind. By the time archaeologists revisited the site in 1985, the markers from 1983 were missing, likely sheared off by trawl gear given the trawling activity in the area and presence of trawl nets hung up on portions of the site (Green et al., 1986). The excavators also noted there were no complete ceramic vessels found, and 'thought that the trawlers have scoured the surface of the site and caused material to be damaged or moved off the site' (Green et al., 1986, p. 116).

In Vietnam, less than two miles from the southern tip of Phu Quoc Island, a team of archaeologists were invited by the Vietnam Salvage Company in 1991 to examine a newly discovered shipwreck, later dated to the fourteenth or early fifteenth century (Blake & Flecker, 1994). Much of the hull remained intact and the vessel was surrounded by a large amount of ceramic, nearly all broken. The archaeologists, Warren Blake and Michael Flecker, noted that 'trawling and the use of explosives, both common fishing methods in the area, could explain the widespread field of broken pottery around the hull' (Blake & Flecker, 1994, p. 73). Additionally, off the coast of NSW Australia, many wrecks have been impacted by scallop dredgers including the *City of Launceston*, *Euralba*, *Eleutheria,* the *Isis*, and the *Lady Darling* (Derksen & Venturoni, 2011).

2.3.1.1 West Africa

The waters around West Africa present an interesting case of fisheries management, foreign exploitation of local livelihoods, and an illustration of the lack of archaeological work in the developing world. Since the 1950s, West Africa's fisheries have been targeted by foreign nations including Russia, Europe, and most recently China (Steadman et al. 2021). In the 1980s, as their own fish stocks showed signs of depletion, China developed a fleet to fish in foreign waters and now deploys trawlers in the EEZs of nearly every country in West Africa. The technologically superior foreign vessels can catch five times as much in one day as a small village fleet gather in one year (Wester, 2023).

This exploitation by foreign vessels has led to conflict between the industrial vessels and local small-scale boats where more than 250 of West African fishers die each year because of collisions or incidents with trawling vessels (Steadman et al. 2021). Journalists have investigated the issue (e.g., Wester, 2023; Jacobs, 2017) and highlight many cases of witnessed corruption within the African fishery inspectors and port authorities. Illegal fishing is also rampant with many vessels catching more than their quotas or switching off their vessel tracking systems (Welch et al., 2022).

The risks to UCH are understudied in this part of the ocean as well. Maritime archaeology in Africa is less common when compared with places like the Mediterranean, European waters, or North America. For example, Gregory Cook, Rachel Horlings, and Andrew Pietruszka's work (Cook, 2012; Cook et al., 2016; Horlings & Cook, 2017) off the coast near Elmina Castle, Ghana was the first in the area and other major maritime archaeology work includes work in Senegal (Guérout,

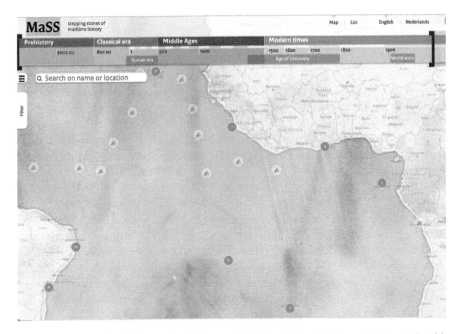

Fig. 2.5 An example of shipwrecks off the coast of Africa. This only shows Dutch vessels with historic evidence for wrecking or archaeological confirmation of a vessel covering centuries of history. Vessels like these ones are at risk from trawling. (Source: Maritime Stepping Stones (MaSS), licensed under the Creative Commons Attribution-ShareAlike license (CC BY-SA))

1996) and Cape Verde (Smith, 2002). The potential that maritime archaeology in West Africa is immense and could be impacted by trawling if it has not been already (Fig. 2.5).

2.3.2 Arctic

UCH in the Arctic is also at risk from mobile fishing gear. A report commissioned by the Nordic Council of Ministers (Ragnarsson et al., 2006) gives an excellent overview of both the ecological damage as well as destruction of UCH presented by Dag Nœvstad. For example, when a vessel was doing maintenance work on an oil platform in the North Sea, an 11-m-long, 2-ton, piece of a ship's keel was found, tangled in a net. It had, seemingly, been dragged by a trawling vessel until being brought to the oil platform near Norway (Nœvstad, 2006). Plenty of prehistoric material, including Stone Age tools, have also been trapped in fishing gear and removed from their contexts (Nœvstad, 2006). Parts of wooden ships dating back to the sixteenth century have also been brought up by trawlers near Greenland (Nœvstad, 2006).

2.3.3 Prehistoric Material

It is not just shipwrecks being destroyed. Among the sites worst affected by trawlers is Doggerland, a 30,000-km^2 area inhabited during the Mesolithic period (8000 years ago), which is now under the waters of the North Sea. Hundreds of stone tools and bones have been dragged up over the centuries as it has been continually targeted by trawlers (Louwe Kooijmans, 1970; Ward & Larcombe, 2008; Mol et al., 2006). This area is a valuable example of a site that is both a paleo-seascape as well as a site of past human inhabitation and rising sea levels have put many other prehistoric landscapes under coastal waters.

Also in the North Sea, near Brown Bank, Dutch fishers found items dating to 7200–6000 BC and near Colijnsplaat, Netherlands, when Roman alter stones were dragged up (Louwe Kooijmans, 1970; Hassal, 1978). Even a Neanderthal skull fragment has been found in this area (Hublin et al., 2009) and many more examples of prehistoric material, such as mammoth bone, have been brought up by North Sea fishers since at least 1874 (Glimmerveen et al., 2008). English fishers have pulled material as well, with notable studies in tools taken by oyster dredgers in Solent, near the Isle of Wight (Momber et al., 2011) and overall interactions between fishing and the North Sea's historic environment (Firth et al., 2013).

There are many examples of this in North America too. As a result of deglaciation and isostatic uplift, many indigenous sites are now covered by water, many are located kilometres offshore. For example, in 1988 a hydraulic clam dredge on Maryland's eastern shore of the Chesapeake found the Nicolas Point site (Lowery & Martin, 2009). The dredges in the area act as a 'backhoe' and 'virtually every clam dredger has discovered prehistoric artefacts while dredging' (Lowery & Martin, 2009, p. 160). At Nicolas Point, this included a rare Archaic-age burial feature. Additionally, fishers off the coast of Maine have found 9000-year-old tools, which prompted an Institute of Maritime History survey sponsored by a NOAA grant.[1]

2.4 Management of Trawling and Fishery Sustainability

Fisheries require management at multiple levels and bottom trawling is no different. Within a nation's Exclusive Economic Zone (200 nautical miles from the shoreline) the coastal state has the rights to fish and responsibility for management. Domestic regulatory framework is the primary way this is done. Within the European Union (EU), the Common Fisheries Policy (CFP) is in place and is managed by the European Commission. At an international level, in the UN fisheries are managed by the Food and Agriculture Organisation (FAO) and Straddling Fish Stocks

---

[1] https://www.maritimehistory.org/submerged-prehistoric-landscape-survey-in-blue-hill-bay-maine/

Agreement (the United National Fish Stock Agreement adopted 1995 and in force in 2001) regulates fish stocks which pass through multiple countries' EEZs. Management and regulation of fisheries continues to advance. As European waters are some most heavily trawled, in 2019, the European Commission implemented 'a global management strategy for the whole western Mediterranean' although each country still manages its own waters and governs which areas or seasons bottom fishing is restricted (Vigo et al., 2023, p. 2). In other waters, however, such as southeast Asia, trawling remains largely unregulated and non-selective (Hilborn et al., 2023).

While bottom trawling is the largest type of physical disturbance by humans to the marine environment, it is also an essential component of the global food supply (Hiddink et al., 2020). The 'magnitude of the effect of the trawl disturbance on benthic communities depends on the frequency of trawling, the impact per trawl pass, and the individual recovery rates of biota exposed to trawling' (Hilborn et al., 2023, p. 568; Hiddink et al., 2017). However, the idea that trawling needs to be banned entirely and is unsustainable is not accurate. Effective management can lead to sustainable fisheries. While many fish stocks worldwide are overexploited, this is a 'failure of fisheries management to control fishing pressure rather than a direct consequence of the fishing gear used' (Hilborn et al., 2023, p. 1568). In essence, bottom trawling in some form is necessary, as it provides a form of food production, which in fact has been shown to have a lesser footprint than other forms, such as aquaculture, crops, and livestock (Hilborn et al., 2023). That having been said, bottom trawls are the least fuel-efficient types of fishing gear, but allowing fish stocks to rebound and be fished sustainably would minimise the distance fishing vessels need to travel to find new grounds.

In addition to fisheries management and sustainability, another recent concern has been the contribution of bottom trawl fishing operations to carbon emissions. One of the major impacts to the seabed of mobile fishing gear is the resuspension of the top strata of sediments that plows through benthic communities and alters biogeochemical processes at the sediment-water interface. The concern is that the resuspension of sediments allows for more mixing of the sediments with seawater and allow remineralisation of carbon initially buried in the seabed to be exposed again to oxygen and prevent its sequestration (Zhang et al., 2023). Initially proposed by Sala et al. (2022), the idea is controversial, as other scholars suggest the increase in nutrients from the resuspended sediments that thereby increase primary production may offset the influx of carbon. However, not to diminish the profound impact of trawling to the marine environment, an evaluation of the carbon flux from sediments due to trawling indicates that only a fraction of sequestered carbon would impact atmospheric CO_2 levels and that there is little evidence trawling has contributed directly to greenhouse gas emissions (Hilborn et al., 2023).

Nevertheless, bottom trawling must be regulated to maintain sustainable fisheries worldwide, which is a trade-off between the detrimental environmental effects with food scarcity, income, and employment (Hilborn et al., 2023). Gear modifications to lessen impacts to the seabed and minimise bycatch of protected species or juveniles can do much to this effect. In addition, management and regulation that reduce the footprint of trawling has shown that this results in higher fish stock yields than if fishing operations are spread over a wider area of seabed (Bloor et al., 2021).

2.5 Marine Protected Areas and Spillover

Hilborn and colleagues (2023) note that sensitive habitats, such as coral reefs or nearshore nurseries, can be protected from fishing activities effectively when locations are known and closed off to vessels, 'prior to significant disturbance' (p. 1573). The same is true for non-natural hard substrate, such as shipwrecks and artificial reefs, which provide important habitat for juvenile fish and concentrate populations. For example, vessel-reefs off southeastern Florida were shown to support a higher species richness and abundance than natural reefs and 'enhanced local fish populations' (Ross et al., 2016, p. 46). If marine protected areas can be established around shipwreck sites, or areas of numerous wrecks, such as offshore certain historic harbours, they can be protected like other areas of sensitive habitat are. This creates a win-win scenario for both marine environmental protection and cultural heritage protection.

An additional benefit to fisheries management and ecosystem protection comes with the concept of 'spillover' where protected areas can allow for populations, especially juveniles, a safe haven to develop, and those populations would then spillover into fishable ground, increasing the fishery. This was put forth for ancient shipwrecks in the Mediterranean area as a management option if marine protected areas were set up around shipwreck sites, juvenile fish populations could thrive and spillover into other areas, thereby protecting the shipwreck sites while helping to increase the fishery (Krumholz & Brennan, 2015). Recent research on seabed recovery from trawl damage has also put forth this argument: 'The establishment of Marine Protected Areas, such as legally recognized no-take reserves where fishing activity is prohibited, could be a useful management measure… the benefits obtained from MPAs could also be observed in adjacent areas, as a result of the spillover of adults and juveniles from the protected area' (Vigo et al., 2023, p. 2). Management synergy can be found here between marine environmentalists, fisheries management, and maritime archaeologists. Developing sustainable fishing regulations and promoting those operations, with less destructive gear, for example, can 'offer de facto protection for UCH' (Pearson & Thompson, 2023), while establishing protections for UCH can also help sustain and increase fish stocks.

Commercial-scale fisheries exist due to the demand for seafood worldwide. Technological advances in fishing gear and vessels continues to drive the industry to put pressure on marine ecosystems (Clare et al., 2023). Marine protected areas and other no-take type of areas are essential for overfished areas to rebound and enhance fishery yields. Maximising mutual benefits of multi-use areas can help 'minimize trade-offs between conflicting preferences' (Clare et al., 2023, p. 1297). Synergy in cultural resources allow shipwrecks to act as artificial reefs and obtain protections themselves within no-fish zones around them. Wrecks have been studied as artificial reefs. The typically vertical structures may not imitate a natural environment but have been shown to 'establish their own community, which is influenced by the spatial orientation and complexity of the structure' (Fagundes-Netto et al., 2011, p. 104). Research off Brazil on the metal-hulled wreck of Orion indicate an increase of juvenile fish around the wreck, supporting the spillover concept (Fagundes-Netto et al., 2011). The trouble with establishing marine protected areas around shipwrecks for their protection and for establishment of protected environments is locating shipwrecks, especially those in deep water, which is a main challenge for shipwreck preservation.

References

Atkinson, C. M. (2012). *Impacts of bottom trawling on underwater cultural heritage* [Master's thesis]. Texas A&M University.

Beare, D., Hölker, F., Engelhard, G. H., McKenzie, E., & Reid, D. G. (2010). An unintended experiment in fisheries science: A marine area protected by war results in Mexican waves in fish numbers-at-age. *Naturwissenschaften, 97*(9), 797–808. https://doi.org/10.1007/s00114-010-0696-5

Blake, W., & Flecker, M. (1994). A preliminary survey of a South-East Asian wreck, Phu Quoc Island, Vietnam. *The International Journal of Nautical Archaeology, 23*(2), 73–91.

Bloor, I. S., Duncan, P. F., Dignan, S. P., Emmerson, J., Beard, D., Gell, F. R., McHarg, K., et al. (2021). Boom not bust: Cooperative management as a mechanism for improving the commercial efficiency and environmental outcomes of regional scallop fisheries. *Marine Policy, 132*, 104649.

Bolster, W. J. (2012). *The mortal sea: Fishing the Atlantic in the age of sail*. Belknap Press of Harvard University Press.

Brennan, M. (2016). Quantifying impacts of trawling to shipwrecks. In M. E. Keith (Ed.), *Site formation processes of submerged shipwrecks* (pp. 157–179). University Press of Florida.

Brennan, M. L., Ballard, R. D., Roman, C., Bell, K. L. C., Buxton, B., Coleman, D. F., Inglis, G., Köyağasıoğlu, O., & Turanlı, T. (2012). Evaluation of the modern submarine landscape off southwestern Turkey through the documentation of ancient shipwreck sites. *Continental Shelf Research, 43*, 55–70. https://doi.org/10.1016/j.csr.2012.04.017

Brennan, M. L., Davis, D., Ballard, R. D., Trembanis, A. C., Vaughn, J. I., Krumholz, J. S., Delgado, J. P., Roman, C. N., Smart, C., Bell, K. L. C., Duman, M., & DuVal, C. (2016). Quantification of bottom trawl fishing damage to ancient shipwreck sites. *Marine Geology, 371*, 82–88. https://doi.org/10.1016/j.margeo.2015.11.001

Brennan, M. L., Delgado, J. P., Jozsef, A., Marx, D. E., & Bierwagen, M. (2023). Site formation processes and pollution risk mitigation of World War II oil tanker shipwrecks: Coimbra and Munger T. Ball. *Journal of Maritime Archaeology, 18*(2), 321–335. https://doi.org/10.1007/s11457-023-09365-4

Bureau of Fisheries. (1914). *Otter-trawl fishery* (App. 6; report for the fiscal year 1914).

Caddy, J. F. (1973). Underwater observations on tracks of dredges and trawls and some effects of dredging on a scallop ground. *Journal of the Fisheries Research Board of Canada, 30*(2), 173–180. https://doi.org/10.1139/f73-032

Christensen, V., Guenette, S., Heymans, J. J., Walters, C. J., Watson, R., Zeller, D., & Pauly, D. (2003). Hundred-year decline of North Atlantic predatory fishes. *Fish and Fisheries, 4*(1), 1–24. https://doi.org/10.1046/j.1467-2979.2003.00103.x

Clare, D. S., Martinez, R., Goodsir, F., Spence, M. A., Tyllinakis, E., Ware, S., & Schratzberger, M. (2023). An approach for assessing and ranking fisheries management scenarios in spatially delimited marine areas. *ICES Journal of Marine Science, 80*, 1291–1302.

Cook, G. (2012). *The maritime archaeology of West Africa in the Atlantic world: Investigations at Elmina, Ghana* [PhD dissertation, Syracuse University]. https://surface.syr.edu/ant_etd/99?utm_source=surface.syr.edu%2Fant_etd%2F99&utm_medium=PDF&utm_campaign=PDFCoverPages

Cook, G. D., Horlings†, R., & Pietruszka, A. (2016). Maritime archaeology and the early Atlantic trade: Research at Elmina, Ghana: G. D. Cook et al.: Early Atlantic trade maritime archaeology at Elmina, Ghana. *International Journal of Nautical Archaeology, 45*(2), 370–387. https://doi.org/10.1111/1095-9270.12180

Currie, D., & Parry, G. (1996). Effects of scallop dredging on a soft sediment community: A large-scale experimental study. *Marine Ecology Progress Series, 134*, 131–150. https://doi.org/10.3354/meps134131

Derksen, F., & Venturoni, M. (2011). *Isis 1892–1932*. Maritime Archaeology Association of Victoria. http://home.vicnet.net.au/~maav/isis.htm

Dobbs, D. (2000). *The great Gulf: Fishermen, scientists, and the struggle to revive the world's greatest fishery*. Island Press.

Duplisea, D. E., Jennings, S., Malcolm, S. J., Parker, R., & Sivyer, D. B. (2001). Modelling potential impacts of bottom trawl fisheries on soft sediment biogeochemistry in the North Sea†. *Geochemical Transactions, 2*(1), 112. https://doi.org/10.1186/1467-4866-2-112

Fagundes-Netto, E. B., Gaelzer, L. R., Coutinho, R., & Zalmon, I. R. (2011). Influence of a shipwreck on a nearshore-reef fish assemblages off the coast of Rio de Janeiro, Brazil. *Latin American Journal of Aquatic Research, 39*(1), 103–116.

FAO. (2020). *The state of world fisheries and aquaculture 2020: Sustainability in action*. FAO. https://doi.org/10.4060/ca9229en

Ferrari, B. (1995). *Physical, biological and cultural factors influencing the formation, stabilisation and protection of archaeological deposits in U.K. coastal waters* [PhD thesis]. St. Andrews.

Firth, A., McAleese, L., Anderson, R., Smith, R., & Woodcock, T. (2013). *Fishing and the historic environment* (EH6204). English Heritage.

Flecker, M. (2002). The ethics, politics, and realities of maritime archaeology in Southeast Asia. *The International Journal of Nautical Archaeology, 31*(1), 12–24. https://doi.org/10.1006/ijna.2002.1017

Friedlander, A. M., Boehlert, G. W., Field, M. E., Mason, J. E., Gardner, J. V., & Dartnell, P. (1999). Sidescan-sonar mapping of benthic trawl marks on the shelf and slope off Eureka, California. *Fishery Bulletin, 97*, 786–801.

Glimmerveen, J., Mol, D., Post, K., Reumer, J., van der Plicht, H., de Vos, J., van Geel, B., van Reenen, G., & Pals, J. P. (2008). The North Sea project: The first paleontological, palynological, and archaeological results. In N. C. Flemming (Ed.), *Submarine prehistoric archaeology of the North Sea: Research priorities and collaboration with industry* (pp. 43–52). Archaeology Data Service. https://doi.org/10.5284/1081706

Green, J., Harper, R., & Intakosi, V. (1986). The Ko Si Chang One shipwreck excavation 1983–1985. A progress report. *International Journal of Nautical Archaeology, 15*(2), 105–122. https://doi.org/10.1111/j.1095-9270.1986.tb00559.x

Guérout, M. (1996). Prospections archéologiques sous-marines aux abords de l'île de Gorée (Sénégal). *West African Journal of Archaeology, 26*(2), 99–121.

Hall, S. J., Basford, D. J., & Robertson, M. R. (1990). The impact of hydraulic dredging for razor clams Ensis sp. on an infaunal community. *Netherlands Journal of Sea Research, 27*(1), 119–125. https://doi.org/10.1016/0077-7579(90)90040-N

Hall-Spencer, J. (2000). Scallop dredging has profound, long-term impacts on maerl habitats. *ICES Journal of Marine Science, 57*(5), 1407–1415. https://doi.org/10.1006/jmsc.2000.0918

Hassal, M. (1978). Britain and the Rhine provinces: Epigraphic evidence for Roman trade. In J. du Plat Taylor & H. Cleere (Eds.), *Roman shipping and trade: Britain and the Rhine provinces* (pp. 41–48). Archaeology Data Service. https://doi.org/10.5284/1081680

He, P., Chopin, F., Suuronen, P., Ferro, R. S. T., & Lansley, J. (2021). *Classification and illustrated definition of fishing gears* (No. 672; FAO fisheries and aquaculture technical paper). FAO. https://doi.org/10.4060/cb4966en

Hiddink, J. G., Jennings, S., Sciberras, M., Szostek, C. L., Hughes, K. M., Ellis, N., Rijnsdorp, A. D., McConnaughey, R. A., Mazor, T., Hilborn, R., Collie, J. S., Pitcher, C. R., Amoroso, R. O., Parma, A. M., Suuronen, P., & Kaiser, M. J. (2017). Global analysis of depletion and recovery of seabed biota after bottom trawling disturbance. *Proceedings of the National Academy of Sciences, 114*(31), 8301–8306. https://doi.org/10.1073/pnas.1618858114

Hiddink, J. G., Kaiser, M. J., Sciberras, M., McConnaughey, R. A., Mazor, T., Hilborn, R., Collie, J. S., Pitcher, C. R., Parma, A. M., Suuronen, P., Rijnsdorp, A. D., & Jennings, S. (2020). Selection of indicators for assessing and managing the impacts of bottom trawling on seabed habitats. *Journal of Applied Ecology, 57*, 1199–1209.

Hilborn, R., Amoroso, R., Collie, J., Hiddink, J. G., Kaiser, M. J., Mazor, T., McConnaughey, R. A., Parma, A. M., Pitcher, C. R., Sciberras, M., & Suuronen, P. (2023). Evaluating the sustainability and environmental impacts of trawling compared to other food production systems. *ICES Journal of Marine Science, 80*, 1567–1579.

Holm, P. (2012). World War II and the "great acceleration" of North Atlantic fisheries. *Global Environment, 5*(10), 66–91. https://doi.org/10.3197/ge.2012.051005

Horlings, R., & Cook, G. (2017). The formation of a West African maritime seascape: Atlantic trade, shipwrecks, and formation processes on the coast of Ghana. In A. Caporaso (Ed.), *Formation processes of maritime archaeological landscapes* (pp. 79–111). Springer International Publishing. https://doi.org/10.1007/978-3-319-48787-8_5

Hublin, J.-J., Weston, D., Gunz, P., Richards, M., Roebroeks, W., Glimmerveen, J., & Anthonis, L. (2009). Out of the North Sea: The Zeeland Ridges Neandertal. *Journal of Human Evolution, 57*(6), 777–785. https://doi.org/10.1016/j.jhevol.2009.09.001

Jacobs, A. (2017, April 30). China's appetite pushes fisheries to the brink. *The New York Times*. https://www.nytimes.com/2017/04/30/world/asia/chinas-appetite-pushes-fisheries-to-the-brink.html

Jennings, S., Dinmore, T. A., Duplisea, D. E., Warr, K. J., & Lancaster, J. E. (2001). Trawling disturbance can modify benthic production processes: *Trawling disturbance and production processes*. *Journal of Animal Ecology, 70*(3), 459–475. https://doi.org/10.1046/j.1365-2656.2001.00504.x

Jones, J. B. (1992). Environmental impact of trawling on the seabed: A review. *New Zealand Journal of Marine and Freshwater Research, 26*(1), 59–67. https://doi.org/10.1080/00288330.1992.9516500

Kaiser, M. J. (1997). An estimate of fishing gear disturbance intensities in the Irish Sea: A comparison of beam trawling and scallop dredging. *Aquatic Conservation, 6*, 269–285.

Krumholz, J. S., & Brennan, M. L. (2015). Fishing for common ground: Investigations of the impact of trawling on ancient shipwreck sites uncovers a potential for management synergy. *Marine Policy, 61*, 127–133. https://doi.org/10.1016/j.marpol.2015.07.009

Lotze, H. K. (2005). Radical changes in the Wadden Sea fauna and flora over the last 2,000 years. *Helgoland Marine Research, 59*(1), 71–83. https://doi.org/10.1007/s10152-004-0208-0

Louwe Kooijmans, L. P. (1970). Mesolithic bone and antler implements from the North Sea and from The Netherlands. *Bericht van de Rijksdienst Oudheidkundig Bodemonderzoek, 20–21*, 27–73. https://hdl.handle.net/1887/1086

Lowery, D., & Martin, R. (2009). Archaeology of marine transgression: An inundated middle archaic burial in Chesapeake Bay, Maryland. *Archaeology of Eastern North America, 37*, 159–173.

Mayer, L. M., Schick, D. F., Findlay, R. H., & Rice, D. L. (1991). Effects of commercial dragging on sedimentary organic matter. *Marine Environmental Research, 31*(4), 249–261. https://doi.org/10.1016/0141-1136(91)90015-Z

Mol, D., Post, K., Reumer, J. W. F., Van Der Plicht, J., De Vos, J., Van Geel, B., Van Reenen, G., Pals, J. P., & Glimmerveen, J. (2006). The Eurogeul—First report of the palaeontological, palynological and archaeological investigations of this part of the North Sea. *Quaternary International, 142–143*, 178–185. https://doi.org/10.1016/j.quaint.2005.03.015

Momber, G., Tomalin, D., Scaife, R., Satchell, J., & Gillespie, J. (2011). Mesolithic occupation at Bouldnor Cliff and the submerged prehistoric landscapes of the Solent. *Archaeology Data Service.* https://doi.org/10.5284/1081835

Napper, I. E., Wright, L. S., Barrett, A. C., Parker-Jurd, F. N. F., & Thompson, R. C. (2022). Potential microplastic release from the maritime industry: Abrasion of rope. *Science of the Total Environment, 804*, 150155. https://doi.org/10.1016/j.scitotenv.2021.150155

National Research Council (U.S.) (Ed.). (2002). *Effects of trawling and dredging on seafloor habitat.* National Academy Press.

Nœvstad, D. (2006). Cultural heritage in Arctic waters. In S. A. Ragnarsson, E. Guijarro Garcia, & Nordic Council of Ministers (Eds.). *Bottom trawling and scallop dredging in the Arctic: Impacts of fishing on non-target species, vulnerable habitats and cultural heritage* (pp. 287–335). Nordic Council of Ministers.

Olsgard, F., Schaanning, M. T., Widdicombe, S., Kendall, M. A., & Austen, M. C. (2008). Effects of bottom trawling on ecosystem functioning. *Journal of Experimental Marine Biology and Ecology, 366*(1–2), 123–133. https://doi.org/10.1016/j.jembe.2008.07.036

Pearson, N., & Thompson, B. S. (2023). Saving two fish with one wreck: Maximizing synergies in marine biodiversity conservation and underwater cultural heritage protection. *Marine Policy, 152*, 105613. https://doi.org/10.1016/j.marpol.2023.105613

Pederson, J., & Dorsey, E. (1997, May 30). *Effects of fishing gear on the sea floor of New England.* Northeastern University.

Pilskaln, C. H., Churchill, J. H., & Mayer, L. M. (1998). Resuspension of sediment by bottom trawling in the Gulf of maine and potential geochemical consequences. *Conservation Biology, 12*(6), 1223–1229.

Ragnarsson, S. A., Guijarro Garcia, E., & Nordic Council of Ministers. (2006). *Bottom trawling and scallop dredging in the Arctic: Impacts of fishing on non-target species, vulnerable habitats and cultural heritage.* Nordic Council of Ministers.

Roberts, C. (2008). *The unnatural history of the sea* (1st paperback ed.). Island Press/Shearwater Books.

Roberts, C. (2012). *The ocean of life: The fate of man and the sea.* Viking.

Robertson, P. (2007). *The Sound of Mull archaeological project: (SOMAP) 1994–2005.* University of Michigan Press. https://doi.org/10.30861/9781407301778

Rose, G. A. (2004). Reconciling overfishing and climate change with stock dynamics of Atlantic cod (Gadus *morhua*) over 500 years. *Canadian Journal of Fisheries and Aquatic Sciences, 61*(9), 1553–1557. https://doi.org/10.1139/f04-173

Ross, S. W., Rhode, M., Viada, S., & Mather, R. (2016). Fish species associated with shipwreck and natural hard-bottom habitats from the middle to outer continental shelf of the Middle Atlantic Night near Norfolk Canyon. *Fishery Bulletin, 114*, 45–57.

Sala, A., Damalas, D., Labanchi, L., Martinsohn, J., Moro, F., Sabatella, R., & Notti, E. (2022). Energy audit and carbon footprint in trawl fisheries. *Scientific Data, 9*, 1–20.

Smith, B. S. (2002). An astrolabe from Passa Pau, Cape Verde Islands. *International Journal of Nautical Archaeology, 31*(1), 99–107. https://doi.org/10.1111/j.1095-9270.2002.tb01404.x

Steadman, D., Thomas, J., Villanueva, V., Lewis, F., Pauly, D., Palomares, M. L., Bailly, N., Levine, M., Virdin, J., Rocliffe, S., & Collinson, T. (2021). *New perspectives on an old fishing practice: Scale, context and impacts of bottom trawling.* https://doi.org/10.13140/RG.2.2.12318.41286/1

Steinmetz, J. H. (2010). *Examining mid-Atlantic ocean shipwrecks and commercial fish trawling & dredging* [Master's thesis]. East Carolina University.

Vigo, M., Navarro, J., Aguzzi, J., Bahamon, N., Garcia, J. A., Rotllant, G., Recasens, L., & Company, J. B. (2023). ROV-based monitoring of passive ecological recovery in a deep-sea no-take fishery reserve. *Science of the Total Environment, 883*, 163339.

Ward, I., & Larcombe, P. (2008). Determining the preservation rating of submerged archaeology in the post-glacial southern North Sea: A first-order geomorphological approach. *Environmental Archaeology, 13*(1), 59–83. https://doi.org/10.1179/174963108x279229

Watling, L., & Norse, E. A. (1998). Disturbance of the seabed by mobile fishing gear: A comparison to forest clearcutting. *Conservation Biology, 12*(6), 1180–1197. https://www.jstor.org/stable/2989836

Welch, H., Clavelle, T., White, T. D., Cimino, M. A., Van Osdel, J., Hochberg, T., Kroodsma, D., & Hazen, E. L. (2022). Hot spots of unseen fishing vessels. *Science Advances, 8*(44), eabq2109. https://doi.org/10.1126/sciadv.abq2109

Wester, T. (2023, June 5). 'They are stealing what should be ours': Chinese trawlers are emptying West African fishing grounds. *The Telegraph*. https://www.telegraph.co.uk/global-health/climate-and-people/how-chinese-trawlers-are-emptying-west-african-fishing-grou/

Willmann, R., & Kelleher, K. (2009). *The sunken billions: The economic justification for fisheries reform* (Other agricultural study 47606; agriculture and rural development). World Bank Group. https://documents.worldbank.org/en/publication/documents-reports/documentdetail/656021468176334381/The-sunken-billions-the-economic-justification-for-fisheries-reform

Zhang, W., Porz, L., Yilmaz, R., Kuhlmann, J., Neumann, A., Liu, B., Muller, D., Spiegel, T., Holtappels, M., Ziebarth, N., Taylor, B., Wallmann, K., Kasten, S., Daewel, U., & Schrum, C. (2023). Impact of bottom trawling on long-term carbon sequestration in shelf sea sediments. *Geophysical Research Letters*. https://doi.org/10.22541/essoar.167578408.84551876/v1

Open Access   This chapter is licensed under the terms of the Creative Commons Attribution 4.0 International License (http://creativecommons.org/licenses/by/4.0/), which permits use, sharing, adaptation, distribution and reproduction in any medium or format, as long as you give appropriate credit to the original author(s) and the source, provide a link to the Creative Commons license and indicate if changes were made.

The images or other third party material in this chapter are included in the chapter's Creative Commons license, unless indicated otherwise in a credit line to the material. If material is not included in the chapter's Creative Commons license and your intended use is not permitted by statutory regulation or exceeds the permitted use, you will need to obtain permission directly from the copyright holder.

Chapter 3
The Duty to Protect Our Ocean Heritage from Bottom Trawling

Maria Pena Ermida

Abstract This chapter seeks to provide an overview of the legal framework surrounding the protection of UCH as a part of the Marine Environment within the context of law of the Sea, focusing particularly on the rules regarding bottom trawling.

3.1 Introduction

A book such as this provides a platform to contribute to the discussion of the protection of Underwater Cultural Heritage (UCH) from bottom trawling. The focus of this chapter lies in its collaboration with archaeology, joining forces with the law to frame the protection of UCH within the legal international arena. It will provide an overview of the framework for protecting UCH, as part of the marine environment, from bottom trawling.

This chapter starts with a discussion of the main legal instruments that contain provisions which flesh out a duty to protect cultural heritage, namely, the 1972 Stockholm Declaration and the World Heritage Convention, the Law of the Sea Convention, the 1992 Rio Declaration, the 2001 UNESCO Underwater Cultural Heritage Convention, and fits the new agreement on Biodiversity Beyond National Jurisdiction (BBNJ) in where relevant. The chapter briefly introduces the Precautionary Principle[1] and how it applies to an activity such as bottom trawling. The final part of the chapter will bring all this together, offering a legal path toward protecting UCH from an activity as hazardous as bottom trawling.

---

[1] Also known as the Precautionary Approach in some cases.

M. P. Ermida
Católica Research Centre for the Future of Law, Lisbon, Portugal
e-mail: mermida@ucp.pt

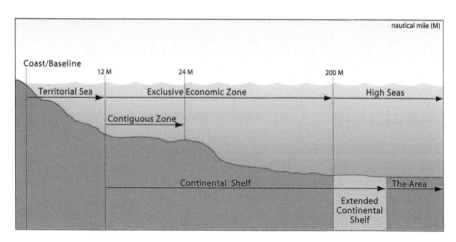

Fig. 3.1 The maritime zones include the 12 nm Territorial Sea, the 24 nm Contiguous Zone, the 200 nm Exclusive Economic Zone, and the High Seas. Image from: United States Department of State. (n.d.). *Maritime zones under the international law of the sea*. https://www.state.gov/about-ecs/

The framework concerning the protection of UCH results from the interaction between different legal entities, namely the law of the sea, environmental law, and cultural heritage. Various legal instruments will be referred to, and although they may seem unrelated at first, the reader will be guided through their unifying points concerning the protection of UCH as a part of the Marine Environment.

The Law of the Sea Convention (LOSC) is the main Convention regarding the ocean, and it strives to protect the global marine water column and the seafloor—regardless of its distance to shore or depth. However, the current governance system for this space is based on an underlying idea of 'divide and conquer', which distinguishes between Areas Within National Jurisdiction and Areas Beyond National Jurisdiction. Areas Within the National Jurisdiction of states include the Territorial Sea (LOSC 1982, Article 2), the Contiguous Zone (LOSC 1982, Article 33), the Exclusive Economic Zone (EEZ), and the Continental Shelf (LOSC 1982, Article 76). Areas beyond National Jurisdiction include the High Seas (the water column and resources therein beyond the EEZ of States) and the Area (the soil, subsoil, and resources beyond the continental shelves of States) (LOSC 1982, Article 133 *et seq.*; LOSC 1982, Part XI) (Fig. 3.1).

3.2 Sources of Duty to Protect Cultural Heritage

3.2.1 The 1972 Stockholm Declaration and the World Heritage Convention

The 1972 Stockholm Declaration (Stockholm Declaration 1973) is usually cited as instrumental in the 'development of provisions that resulted in the 1982 Law of the Sea Convention and the framework for the conduct of activities at sea, which must consider the duties to protect our natural and cultural heritage' (Varmer, 2020, p. 88; Schneider, 1979). It codified the customary practice of nations to balance development with protecting the environment so that future generations may inherit it in a healthy state. The framework contains principles that document and delineate the duty to protect the environment and a duty to cooperate for that purpose under customary international law. For instance, principle 4 mentions a 'special responsibility' to safeguard and wisely manage the heritage of wildlife and its habitat, which is now gravely imperilled by a combination of adverse factors.

Also in 1972, the World Heritage Convention took place and played a significant role in integrating the conservation of natural and cultural heritage. With 193 State parties, it focused initially on terrestrial sites and traditional cultural structures. However, it soon evolved to recognise heritage in the marine environment beyond the territorial sea into the EEZ/CS. This move seaward continues as there are calls for recognition of heritage in the high seas, including wreck sites such as *Titanic*.

Considering these two legal instruments, it is clear how the awareness of the Global Community grew regarding the protection of UCH as a part of the Marine Environment. This elevated UCH makes it a target of protection against human activities capable of causing adverse impacts in the marine environment, such as bottom trawling.

3.2.2 The 1982 UN Law of the Sea Convention

The Law of the Sea Convention, also referred to as the Constitution of the Oceans (Koh, 1982), is the comprehensive legal framework instrument that balances specific legal values such as the flag State rights of navigation, fishing, and other uses with the coastal State jurisdiction, and authority in the maritime zones under its jurisdiction and even, in certain occasions, beyond them. The Convention, however, does not contain any specific mention of bottom trawling. This practice is subject to the general fisheries provisions of the Convention, which do not deal with this activity's particular destructive power.

The LOSC should, however, always be read in line with the implementing agreements that were adopted posteriorly, namely the 1994 Agreement on Part 11, the 1995 Fish Stocks Agreement,[2] and the more recent 2023 Agreement on the Conservation and Sustainable Use of Marine Biological Diversity of Areas Beyond National Jurisdiction (BBNJ). Although the relevance on the 1994 Agreement regarding the issue of bottom trawling is marginal, the same cannot be said about the latter two. While neither specifically refer to bottom trawling, they do contain some relevant provisions for the issue at hand, which will be analysed further in this text. However, both agreements are quite different in how they are structured. The

---

[2] The 1994 Agreement relating to the Implementation of Part XI of the United Nations Convention on the Law of the Sea of 10 December 1982 on part XI and the 1995 Agreement for the Implementation of the Provisions of the United Nations Convention on the Law of the Sea of 10 December 1982 Relating to the Conservation and Management of Straddling Fish Stocks and Highly Migratory Fish Stocks.

1995 Agreement has a purely sectoral approach while the BBNJ attempts to move away from it, addressing ABNJ as a whole, with the goal of 'protecting, caring for and ensuring responsible use of the marine environment, maintaining the integrity of ocean ecosystems and conserving the inherent value of biological diversity of areas beyond national jurisdiction'.[3]

Being a widely accepted international agreement ratified by 168 nations plus the European Union,[4] most of the LOSC is recognised as having codified customary international law (Churchill, 2015, p. 30). As a result, over the past four decades, it has been followed in practice by parties and non-parties, with many rights and obligations that result from this treaty being pointed at as being customary international law (Churchill, 2015, p. 30). This includes a duty to protect UCH.[5]

The LOSC does contain provisions on the 'protection' and 'preservation' of our cultural heritage, referred to in the Convention as 'archaeological and historical objects' (Scovazzi, 2017a, b),[6] in Articles 149 and 303. Article 303 is in the general provisions of Part XVI of the LOSC, while Article 149 regards the Area. (LOSC 1982, Article 1 (1) (1)).[7] These short articles make up a broad and somewhat vague duty to protect UCH without any definitions or further details to understand ways to implement it. Another relevant provision in this regard is Article 192 which focuses on the obligation of the protection of the marine environment. These provisions will be taken as a starting point for this analysis.

3.2.2.1 Duty to Protect Cultural Heritage Under Article 149 and 303

Article 149

According to Article 149 'all objects of an archaeological and historical nature found in the Area shall be preserved or disposed of for the benefit of mankind as a whole, particular regard being paid to the preferential rights of the State or country

---

[3] BBNJ Agreement, Preamble.

[4] 1982 United Nations Convention on the Law of the Sea – Treaty Status. Retrieved September 3 2023 from https://treaties.un.org/pages/ViewDetailsIII.aspx?src=TREATY&mtdsg_no=XXI-6&chapter=21&Temp=mtdsg3&clang=_en#:~:text=Signatories%20%3A%20157.,Parties%20%3A%20169.&text=CTC%2DArabic%3B%20CTC%2DChinese,1833%2C%20p

[5] Take for instance the case of the US, which although being criticised for not ratifying the 2001 UNESCO Convention on the protection of UCH, has on numerous occasions adopted domestic laws that seek to protect UCH.For example, Antiquities Act of 1906, the Archaeological Resources Protection Act of 1979, or the National Marine Sanctuaries Act of 1972, to name a few. The US was a facilitator in the adoption of the Agreement Concerning the Shipwrecked Vessel RMS Titanic 2003 which mentions in its Preamble that UCH should be protected in the interest of present and future generations.

[6] A seemingly deceiving provision which should however be read in a broad sense to include artefacts undoubtedly within the field of archeology but also those of relatively recent origin but that hold a historical weight, such as a sunken ship from WWII. For this reason, throughout this text, for a question of clarity the term used to describe such artefacts will be Underwater Cultural Heritage as defined in the 2001 Convention on Underwater Cultural Heritage (UCH).

[7] The seabed and ocean floor and subsoil thereof, beyond the limits of national jurisdiction.

3 The Duty to Protect Our Ocean Heritage from Bottom Trawling   31

of origin, or the State of cultural origin, or the State of historical and archaeological origin'. This gives UCH similar legal protection to the Area and its resources, which are subject to the principle of the Common Heritage of Humankind (Scovazzi, 2017a; Aznar, 2017). Moreover, the drafters have a clear preference to protect UCH for the public good over the private interest. Although there is no specific mention of the protection of UCH against economic activities, such as bottom trawling, these main ideas are cited as a source of inspiration for the duty to protect UCH contained in the 2001 Convention on the protection of Underwater Cultural Heritage (Scovazzi, 2017a, mn. 15).

Article 303

Article 303(1) establishes two general duties: the duty to protect UCH and the duty to cooperate in doing so (Scovazzi, 2017b, mn.10). For instance, Article 303 (2) limits the geographic scope of coastal State jurisdiction to protect UCH from foreign-flagged vessels and nationals up to the 24-mile limit of the Contiguous Zone. Beyond that, the Coastal state only holds sovereign rights over the 'natural resources'—a concept that does not encompass UCH (Aznar, 2014; Oxman, 1988, p. 363; Scovazzi, 2017b, mn. 20).[8] However, if a foreign national or vessel conducts activities that trigger this jurisdiction over natural resources in the EEZ/CS, namely due to treasure hunters, salvors, or looters, a coastal state may enforce its natural resource regulations against them.

The LOSC also recognises that coastal States have jurisdiction, authority, and control over the placement and management of artificial reefs. Thus, if UCH on the CS beyond the 24 nm Contiguous Zone serves as an artificial reef (as many shipwrecks do), a coastal State may be able to protect, manage, and prevent looting and unwanted salvage under that regime. This aligns with the LOSC drafters' intention to avoid any significant erosion of the principle of freedom of the high seas, particularly regarding the 'creeping jurisdiction' of coastal States in areas beyond the territorial sea (Oxman, 1988, p. 363).

There is, however, no concern for the protection of UCH indirectly harmed by human activities, such as bottom trawling, but activities directed at the pursuit of UCH itself. Nevertheless, it is a starting point to argue that the LOSC *does* establish a duty to protect UCH against human activities, such as bottom trawling, which are directed at exploiting resources around and that can impact UCH.

Moreover, the text in Article 303(4) implicitly recognises that there is little guidance or detail concerning the scope and reach of this duty (Aznar, 2022; Varmer, 2020, p. 77; Scovazzi, 2017b), mn. 10). This provision is therefore an open clause calling for a further expansion in international law of the duties enounced in this

---

[8] The looting in salvage is not within the scope of the rights or freedom of navigation.

provision (Rosenne & Sohn, 1989, p. 162).[9] This need for further details on how to implement this duty to protect cultural heritage has been largely addressed by the UNESCO 2001 Convention discussed below. However, before doing so, it is worth looking at the Duty to Protect the Marine Environment under the LOSC contained in Article 192. This may provide better guidance, particularly when cultural heritage is inextricably integrated with the natural heritage, as it is when shipwrecks act as artificial reefs.

3.2.2.2 Duty to Protect Cultural Heritage Under Article 192

Article 192 of the LOSC establishes a 'general obligation to protect and preserve the marine environment' (LOSC 1982, Article 192). Although the concept of Marine environment is not defined in Article 1 of the LOSC, it could be argued, taking from the preparatory works of the Convention (Malta Draft Articles, 1973), that the drafters sought to go beyond an anthropocentric understanding of the term 'environment' and intended it to include the entire marine ecosystem, especially the habitats of species, many targeted by bottom trawling and in areas which often contain UCH (Czybulka, 2017, mn. 25). Thus, the obligation under Part XII provides guidance on how to address the threat to the marine environment from bottom trawling and may also provide guidance to how to implement the duties under Articles 303 and 149 (Varmer, 2020, p. 92).

3.2.3 The 1992 Rio Declaration

The 1992 Conference on Environment and Development in Rio de Janeiro, Brazil, resulted in the Rio Declaration on Environment and Development, which established in Agenda 21, a duty to protect the marine environment and to cooperate for that purpose—expressly stemming from the 1982 LOSC.[10] One of the most essential aspects of Agenda 21, however, is the fact that it identifies approaches to implement this duty and calls explicitly for integrated management and a precautionary approach to the sustainable development and protection of the marine environment. As Varmer notes, 'While much of the focus is on the conservation of marine living resources, the consideration of cultural heritage can be found throughout, including environmental impact assessments and integrated management' (Varmer, 2020, p. 88).

---

[9] Predicting that the details needed for this new area of international underwater heritage law would be addressed by State practice and presumably in a new UNESCO Convention.

[10] Chapter 17.1 highlights how the LOSC 'sets forth rights and obligations of States and provides the international basis upon which to pursue the protection and sustainable development of the marine and coastal environment and its resources.'

Thus, although it is undeniable that the LOSC provided the first tier of the framework for the legal protection of UCH found in all maritime zones, other tools, put into place since the signing of the LOSC, have been vital to build a structure to protect UCH, not only from salvage and looters but also from activities with the potential of having a negative impact on the marine environment, including bottom trawling.

3.2.4 The UNESCO 2001 Convention

The UNESCO 2001 Convention arose as the Agreement to implement the call for more details under LOSC Art 303(4).[11] At the negotiation meetings, there was consensus regarding four principles: (1) to protect and preserve UCH; (2) the preferred first policy option of *in situ* preservation and adherence to the Annex Rules when a party decides not to preserve *in situ*, and recovery is in the public interest; (3) no 'commercial exploitation' of UCH; and (4) cooperation among States to protect UCH, particularly for training, education, and outreach. The primary purpose and focus were to address the threat from activities directed at UCH, such as looting and unwanted salvage. However, there are some provisions regarding human activities, with the indirect potential of damaging or destroying UCH, such as bottom trawling.

Firstly, the preamble of the Convention highlights 'the need to respond appropriately to the possible negative impact on the underwater cultural heritage of legitimate activities that may incidentally affect it'. Article 2 (3), in turn, emphasises the idea of an obligation of preservation of UCH in the Area, as stated in Article 149 of the LOSC, placing a general duty of protection for all UCH even if there is no direct interest for a State in doing so (Blake, 2015, p. 99). Furthermore, Article 8 of the 2001 Convention provides that 'States Parties may regulate and authorize activities directed at underwater cultural heritage within their contiguous zone', making it clear that there is indeed an obligation to actively protect UCH through coastal state jurisdiction, and also reinforcing the idea that the duty to protect UCH does include economic activities that may adversely impact UCH, such as bottom trawling.

The Annex to the 2001 Convention, concerning activities directed at UCH, also addresses the threats to natural resources and the environment around it, hinting at the idea that UCH is, in fact, a part of the marine environment. For instance, its drafters stated in Rule 10(l) that the project design required for all activities concerning UCH, mentioned in Rule 9, must contain an environmental policy. According to Rule 29, the project design must also be 'adequate to ensure that the seabed and marine life are not unduly disturbed'. This concern is again expressed in Rule 14, which states that 'The preliminary work referred to in Rule 10 (a) shall include an assessment that evaluates the significance and vulnerability of the underwater

---

[11] There are currently 72 parties to the 2001 Convention.

cultural heritage and the surrounding natural environment to damage by the proposed project, and the potential to obtain data that would meet the project objectives'.

Thus, the 2001 Convention provides guidance for implementing the duty to protect under the LOSC, feeding the notion that UCH is often inextricably connected to the natural marine environment, namely when UCH also serves as an artificial reef, thus reinforcing the protection of Article 192 of the LOSC.

3.3 The Precautionary Principle

It is generally agreed upon that applying the precautionary principle presupposes that we are before three cumulative requirements. Firstly, the object of potential application of this must be an activity or substance. Secondly, said object must pose a risk of serious or irreversible harm. Thirdly, there must be scientific uncertainty about the degree, likeliness, or type of damage that can be caused by such an activity or substance (Recuerda, 2008, p. 10).[12] However, this was not always so. In fact, in its early stages, the focus of international environmental law was a reactive one, centring on remedying damage that had already occurred or on preventing damage that was sure to occur (Trail Smelter Case (US v Canada), 1938, p. 1965). Precaution, meaning the idea of seeing beyond the cause-effect paradigm and instead thinking of an unidentified risk or potential future harm, was not a factor that entered the discussion for various decades. This consciousness only arose with the scientific recognition of the delicate balance of ecosystems, the uncertainties surrounding them, and the effects certain activities can have on them and Humankind (Jaeckel, 2017, p. 28).

This has been included in several international legal instruments, such as the 1992 Rio Declaration which is yet another instrument famous for its Principle 15 on the Precautionary Principle and Agenda 21 of specifically, Chapter 17.22, which calls for a precautionary approach to the protection of the marine environment, the 1995 Fish Stocks Agreement,[13] and, more recently, the Agreement on the Protection of Biodiversity Beyond National Jurisdiction, still to enter into force.

International Jurisprudence, mainly that produced by the International Tribunal for the Law of the Sea (ITLOS), has also been an avid defender of the application of a precautionary approach to the protection of the marine environment, including fishing. On ITLOS alone, five occasions can be identified with such an underlying

---

[12] Scientific uncertainty may arise due to a lack of data, the dubious origin of that data or even from contradicting data.

[13] The precautionary approach may also be found in the 1992 United Nations Framework Convention on Climate Change, the 1995 WTO Agreement on the Application of Sanitary and Phytosanitary Measures, the 1998 Agreement on the International Dolphin Conservation Program, the 2000 Convention on the Conservation and Management of Highly Migratory Fish Stocks in the Western and Central Pacific Ocean, the 2001 International Convention on the Control of Harmful Anti-Fouling Systems, and the 2001 Agreement on the Conservation of Albatrosses, to name a few.

idea: three court cases, namely, the Southern Bluefin Tuna, the MOX Plant, and the Land Reclamation cases, and two Advisory Opinions, specifically that on state Responsibility concerning activities in the Area and that concerning Illegal, Unreported, and Unregulated (IUU) Fishing.

Although the claimants mention the precautionary approach in all three court cases in their arguments, the ITLOS did not phrase it as such. Instead, the court kept to less compromising language such as 'for instance' and the use of expressions such as 'prudence' and 'caution' (Southern Bluefin Tuna, *(New Zealand v. Japan; Australia v. Japan)* 1999, para. [77]; MOX Plant (Ireland v. United Kingdom), 2001, para [84]; Land Reclamation, (Malaysia v Singapore) 2003, para [99]). The ITLOS Advisory Opinions, however, reveal an evolution in this sense, referring clearly to a 'precautionary approach' and even evoking Principle 15 of the Rio Declaration (AO ITLOS, 2011, para. [127]). In fact, in its 2015 Advisory Opinion, in addressing the issue concerning flag State obligations in the case of IUU Fishing, the Tribunal already mentioned that parties should apply conservation and management measures based on the best available scientific evidence but that if said evidence is insufficient, the 'precautionary approach' should be applied (AO ITLOS 2015, para. [208]).

As Kristina Gjerde beautifully puts it, 'to ensure consistent application of principles agreed to by the world community (…), the principles, policies, and best practices that were adopted (…) will need to be explicitly recognized and incorporated into management action at all levels' (Gjerde, 2006, p. 305). Thus, a path emerges where the precautionary principle is included into both legal instruments as well as case law. Potentially, it could even become a common standard that applies to fishing activities, including bottom trawling (Sands et al., 2012, p. 225).

3.4 Steps Forward

The last few years have seen an increase in global concern over the impact fisheries, and in particular bottom trawling, on the marine environment, particularly in what concerns vulnerable marine ecosystems (VME). However, there has been little political will to take real international action against such threats. For instance, when United Nations General Assembly (UNGA) attempted to move towards a ban of bottom trawling in 2006, the measures were blocked (Goodwyn, 2015, p. 808). However, in 2007 and 2009, the UNGA adopted Resolution 61/105 and 64/72. In both resolutions, there is a clear emphasis on a need for sustainable fisheries. In addition, these resolutions also included a recognition of the work that was already carried out by Regional Fisheries Management Organisations (RFMOs),[14] in what concerns bottom trawling. The latest 2011 UNGA resolution 66/68 led to a further

---

[14] International organisations meant to regulate regional fishing activities in the high seas. For example, North-East Atlantic Fisheries Commission or the International Commission for the Conservation of Atlantic Tunas.

increase in management bodies' efforts concerning the establishment of closed areas to protect VMEs. These resolutions have sustained a call to action for these entities to further adopt and start to 'implement measures, in accordance with the precautionary approach, ecosystem approaches, and international law, for their respective regulatory areas as a matter of priority' (UN, 2006, Para. [83]). These measures have been mainly of two types: Area Based Management Tools (ABMT) and rules concerning fishing gear.

Regarding fishing gear, in 2009, FAO released the International Guidelines for the Management of Deep-Sea Fisheries in the High Seas (FAO, 2008). Bottom fishing gear has been shown to impact wrecks severely and has likely caused oil release events due to damage from mobile gear (e.g., Brennan, 2016; Brennan et al., 2023). Although this document only targeted specific VME and biodiversity protection,[15] it has contributed to changes in the industry, namely through the modification of gear to reduce the effects of this activity in the deep seabed, namely by reducing its weight and size or building it in a way that keeps most of it above the seabed (O'Leary et al., 2020, p. 5).

Nevertheless, it is generally agreed upon that the most effective option to mitigate human impact within the marine environment are ABMT (O'Leary et al., 2020, p. 9). ABMT aim to manage human activities, establish areas for biodiversity and ecosystem conservation, and sustainable use of resources. RFMOs have been essential in this regard, adopting measures to control bottom trawling, namely through MPAs (Rayfuse, 2015, p. 1296).

One of the ways in which the BBNJ agreement proposes to do this is through the establishment of ABMT.[16] This is to be done through a framework based on cooperation and coordination as the BBNJ attempts to strike a balance between the interests of the Agreement and those of all individual Parties and established international bodies.[17] There is undeniable potential for adopting solutions and enhanced cooperation in protecting the Marine Environment, particularly UCH, in a legal framework such as the BBNJ. However, at the present moment, there are various uncertainties as to when this agreement and its institutions will enter into force, if at all (Chavez-Molina et al., 2023).

Nevertheless, we can confidently say there is a light at the end of the tunnel. The Vienna Convention on the Law of the Treaties contains certain obligations regarding

---

[15] For instance, according to Chapter 5.2 'A marine ecosystem should be classified as vulnerable based on the characteristics that it possesses'. The guidelines do provide a list of such characteristics which includes uniqueness or rarity, functional significance of the habitat, fragility, life-history traits of component species that make recovery difficult and structural complexity. Examples contained in Annex I of the guidelines include 'Coldwater corals and hydroids', 'some types of sponge dominated communities', 'communities composed of dense emergent fauna where large sessile protozoans (xenophyophores) and invertebrates (e.g., hydroids and bryozoans) form an important structural component of habitat' and endemic 'seep and vent communities comprised of invertebrate and microbial species'.

[16] BBNJ Agreement, Article 1(1), BBNJ Agreement, Article 17 (a), (b) and (c).

[17] BBNJ Agreement, Article 17 (d), (e) and Article 19 (2) and (3).

the actions of States that may defeat the object and purpose of a Treaty,[18] which in this case include ensuring the conservation of marine biodiversity in areas beyond national jurisdiction.[19] Moreover, the legal duties to protect our Ocean Heritage under international law and the goals of the UN Decade of Ocean Science for Sustainable Development (UN, 2017) show that the best way to address the threats from bottom trawling is a precautionary approach. This approach has gained traction within legal instruments and decisions in the last decade and has even been included in the final text of the BBNJ. The application of precautionary measures such as a temporary moratorium regarding bottom trawling until several conditions are met, namely surveys to ensure that no UCH finds itself in planned exploitation areas, proper Environmental Impact Assessments, and significant natural and cultural heritage aside as Marine Protected Areas. This is the logical solution against a fishing technique that can result in irreparable harm and destruction to UCH and marine life. Suppose this precautionary principle is respected and the standards are developed and followed. In that case, the results will be a healthier ocean, a necessary characteristic of Earth's life support system, as well as the safeguarding of humanity's own culture and history through the preservation of UCH.

References

Aznar, M. J. (2014). The contiguous zone as archaeological zone. *The International Journal of Marine and Coastal Law, 29*, 1–51.

Aznar, M. J. (2017). Exporting environmental standards to the protection of underwater cultural heritage in the area. In J. Crawford, A. G. Koroma, S. Mahmoudi, & A. Pellet (Eds.), *The international legal order: Current needs and possible responses – Essays in honour Djamchid Momtaz* (1st ed., pp. 255–273). Brill/Nijhoff.

Aznar, M. J. (2022). Comentário ao artigo 303° objetos arqueológicos e históricos achados no mar. In W. Brito (Ed.), *Comentários à Convenção das Nações Unidas sobre Direito do Mar* (1st ed., pp. 567–593). Ed. Almedina.

Blake, J. (2015). *International cultural heritage law*. Oxford University Press.

Brennan, M. (2016). Quantifying impacts of trawling to shipwrecks. In M. E. Keith (Ed.), *Site formation processes of submerged shipwrecks* (pp. 157–179). University Press of Florida.

Brennan, M. L., Delgado, J. P., Jozsef, A., Marx, D. E., & Bierwagen, M. (2023). *Site formation processes and pollution risk mitigation of World War II oil tanker shipwrecks: Coimbra and Munger T. Ball* [Preprint]. In Review. https://doi.org/10.21203/rs.3.rs-3029829/v1

Chavez-Molina, V., et al. (2023). Protecting the Salas y Gomez and Nazca Ridges: A review of policy pathways for creating conservation measures in the international waters of the Southeast Pacific. *Marine Policy, 152*, 105594.

Churchill, R. R. (2015). The United Nations Convention on the Law of the Sea. In D. R. Rothwell, A. G. O. Elferink, K. N. Scott, & T. Stephens (Eds.), *The Oxford handbook on law of the sea* (1st ed., pp. 24–45). Oxford University Press.

---

[18] Vienna Convention on the Law of the Treaties, Article 18, Article 30.

[19] BBNJ Agreement, Article 2.

Czybulka. (2017). Art. 192. In A. Proelss (Ed.), Maggio A. R. (Assistant Editors), Blitza E. (Assistant Editors), Daum O. (Assistant Editors), *United Nations Convention on the Law of the Sea: A commentary*. Nomos Verlagsgesellschaft.

Food and Agriculture Organisation (FAO). (2008, August 29). *International guidelines for the management of deep-sea fisheries in the high seas*. SPRFMO-VI-SWG-INF01. Retrieved August 2, 2023, from https://www.sprfmo.int/assets/Meetings/Meetings-before-2013/Scientific-Working-Group/SWG-06-2008/7e88751513/SPRFMO6-SWG-INF01-FAO-Deepwater-Guidelines-Final-Sep20.pdf

Gjerde, K. (2006). High seas fisheries management under the convention on the law of the sea. In D. Freestone, R. Barnes, & D. M. Ong (Eds.), *The law of the sea: Progress and prospects* (1st ed., pp. 281–307). Cambridge University Press.

Goodwyn, E. (2015). Threatened species and vulnerable marine ecosystems. In D. R. Rothwell, A. G. O. Elferink, K. N. Scott, & T. Stephens (Eds.), *The Oxford handbook on law of the sea* (1st ed., pp. 799–824). Oxford University Press.

Jaeckel, A. L. (2017). *The international seabed authority and the precautionary principle*. Brill|Nijhoff.

Koh, T. B. (1982, December 6 and 11). A constitution for the oceans. Montego Bay conference. Retrieved August 2023, from https://www.un.org/depts/los/convention_agreements/texts/koh_english.pdf

O'Leary, B. C., et al. (2020). Options for managing human threats to high seas biodiversity. *Ocean and Coastal Management, 187*, 105110.

Oxman, B. (1988). Marine archaeology and the international law of the sea. *Columbia Journal of Law, 12*(353), 353–372.

Rayfuse, R. (2015). Regional fisheries management organizations. In D. R. Rothwell, A. G. O. Elferink, K. N. Scott, & T. Stephens (Eds.), *The Oxford handbook on law of the sea* (1st ed., pp. 349–462). Oxford University Press.

Recuerda, M. A. (2008). Dangerous interpretations of the precautionary principle and the foundational values of European Union food law: Risk versus risk. *Journal of Food and Law Policy, 4*(1), 1–42.

Rosenne, S., & Sohn, L. (Eds.). (1989). *United Nations law of the sea convention 1982: A commentary: Volume V*. Martinus Nijhof.

Sands, P., Peel, J., Fabra, A., & MacKenzie, R. (2012). *Principles of international environmental law* (3rd ed.). Cambridge University Press.

Schneider, J. (1979). *World public order of the environment: Towards an international ecological law and organization*. University of Toronto Press, Scholarly Publishing.

Scovazzi, T. (2017a). Article 149. In A. Proelss (Ed.), A. R. Maggio (Assistant Editors), E. Blitza (Assistant Editors), O. Daum (Assistant Editors) (2017). *United Nations Convention on the Law of the Sea: A commentary*. Nomos Verlagsgesellschaft.

Scovazzi, T. (2017b). Article 303. In A. Proelss (Ed.), A. R. Maggio (Assistant Editors), E. Blitza (Assistant Editors), O. Daum (Assistant Editors) (2017). *United Nations Convention on the Law of the Sea: A commentary*. Nomos Verlagsgesellschaft.

United Nations General Assembly. (2006, November 8). Sustainable fisheries, including through the 1995 agreement for the implementation of the provisions of the United Nations Convention on the Law of the Sea of 10 December 1982 relating to the Conservation and Management of Straddling Fish Stocks and Highly Migratory Fish Stocks, and related instruments. A/RES/61/105 Retrieved September 2, 2023, from https://digitallibrary.un.org/record/588357?ln=en

United Nations General Assembly. (2017, November 22). *Seventy-second session Agenda item 77 (a): Oceans and the law of the sea: oceans and the law of the sea*. Resolution A/72/L.18. Retrieved August 2, 2023, from https://documents-dds-ny.un.org/doc/UNDOC/LTD/N17/396/23/PDF/N1739623.pdf?OpenElement

Varmer, O. (2020). The duty to protect underwater cultural heritage and to cooperate for that purpose under law and policy. In H. Karan, X. Ademuni-Odeke, & K. Türk (Eds.), *The legal regime of underwater cultural heritage and marine scientific research* (1st ed., pp. 77–117). Yetkin Yayinlari.

Cases

Land Reclamation in and around the Straits of Johor (Malaysia v. Singapore), Provisional Measures, Order of 8 October 2003, ITLOS Reports 2003, p. 10.

MOX Plant (Ireland v. United Kingdom), Provisional Measures, Order of 3 December 2001, ITLOS Reports 2001, p. 95.

Request for Advisory Opinion submitted by the Sub-Regional Fisheries Commission, Advisory Opinion, 2 April 2015, ITLOS Reports 2015, p. 4.

Responsibilities and obligations of States with respect to activities in the Area, Advisory Opinion, 1 February 2011, ITLOS Reports 2011, p. 10.

Southern Bluefin Tuna (New Zealand v. Japan; Australia v. Japan), Provisional Measures, Order of 27 August 1999, ITLOS Reports 1999, p. 280.

Trail Smelter Case (US v Canada), Arbitration Award, 16 April 1938 and 11 March 1941, 3 RIAA 1905.

Open Access This chapter is licensed under the terms of the Creative Commons Attribution 4.0 International License (http://creativecommons.org/licenses/by/4.0/), which permits use, sharing, adaptation, distribution and reproduction in any medium or format, as long as you give appropriate credit to the original author(s) and the source, provide a link to the Creative Commons license and indicate if changes were made.

The images or other third party material in this chapter are included in the chapter's Creative Commons license, unless indicated otherwise in a credit line to the material. If material is not included in the chapter's Creative Commons license and your intended use is not permitted by statutory regulation or exceeds the permitted use, you will need to obtain permission directly from the copyright holder.

Chapter 4
Quantification of Bottom Trawl Damage to Ancient Shipwrecks: A Case Study from the Coastal Waters of Turkey

Michael L. Brennan

Ancient shipwrecks in the Mediterranean region are particularly susceptible to damage by bottom trawl fishing activities because of their low relief on the seabed. Shipwrecks from the Greek and Roman periods typically consist of mounds of amphora cargoes, which slump onto the seabed in the shape of the vessel outline once the wooden hulls deteriorate following consumption by wood boring organisms. These low relief mounds do not snag trawl nets the way steel-hulled shipwrecks do in which fishers lose gear and could avoid those locations later. Instead, the fragile ceramic artefacts that comprise ancient wrecks are exposed to repeated strikes by trawls until they are broken and scattered across the seabed, potentially to the extent that they can no longer be found. Expeditions from 2008 through 2013 by Ocean Exploration Trust (OET) and the Exploration Vessel *Nautilus* documented numerous ancient shipwrecks in the Black and Aegean Sea coastal waters of Turkey that exhibited a range of damage from trawls, and which allowed for quantification of some of this damage.

Understanding the effects of bottom trawling on ancient wrecks, and assessing the threat posed to them, requires finding the wrecks, documenting their current state of preservation, and finally quantifying the amount of damage already inflicted, prior to the implementation of any protection measures. Hard bottom substrate, which the amphora cargoes of these wrecks provide, are important habitat for juvenile fish and other organisms as artificial reefs, and trawl activities cause severe impacts to the benthos. The drastic effects of continued trawl activities also include the smoothing over of seabed substrates and erasure of seabed structures such as ripples and other bedforms, soft reef substrate like carbonate outcrops, and other structures such as wrecks (Vigo et al., 2023; Puig et al., 2012). Efforts to protect shipwreck sites specifically from trawling have been minimal. For shipwrecks in

---

M. L. Brennan (✉)
SEARCH Inc., Jacksonville, FL, USA
e-mail: mike@brennanexploration.com

© The Author(s) 2024
C. Jarvis (ed.), *Threats to Our Ocean Heritage: Bottom Trawling*,
SpringerBriefs in Underwater Archaeology, https://doi.org/10.1007/978-3-031-57953-0_4

deep water, and unobservable to researchers and resource managers, the essential component missing in reducing the damage inflicted by bottom trawling is thorough documentation of threatened sites by return visits and repeated surveys (Brennan et al., 2016).

4.1 Trawling in the Mediterranean

The Mediterranean Sea is home to the oldest – and some of the most overexploited – trawling grounds in the world. The deep-sea fishery was historically centred on the deep-water red shrimp species, *Aristeus antennatus* and *Aristaeomorpha foliacea*, originating in the western and central Mediterranean. When the eastern Mediterranean countries, namely Greece and Turkey, began to exploit their own deep-water resources in the early 2000s, the fishing grounds were determined as non-virginal seabed, since Italian trawlers had already been exploiting them (Pulcinella et al., 2023; Pinello et al., 2018). Over the last two decades, the coastal resources of the Mediterranean have become overexploited due to the increasing power of trawling vessels and the expansion of benthic fisheries deeper onto the continental shelf down to depths of 1000 m. Recent work has investigated the 'spatiotemporal patterns of trawling pressure and impact… particularly towards commercial species and unwanted catches, but also as trawling extent and intensity on European wide habitats' (Smith et al., 2023, p. 2). This chapter builds upon previous history of Turkish fisheries, specifically, presented in Brennan et al. (2012), to update the discussion with recent scholarship.

European Union guidance has begun to further regulate the fishing industry with the objective of assuring that the marine waters of the Mediterranean Sea are in 'Good Environmental Status (GES)' (Smith et al., 2023, p. 1) to help determine, and improve, the pressures and impacts of fisheries on the marine ecosystem and provide better ecosystem management. Other major efforts in Europe in recent years have included the BENTHIS project[1] which conducts mapping of the seabed for the purposes of identifying trawling impacts (Eigaard et al., 2017). Such work has produced a series of scientific papers that discuss large areas of the Mediterranean and document trawling impacts across the entire sea 'for benthic status in relation to trawling intensity, landings and value' for habitats at varying depths (Smith et al., 2023, p. 2). In 2019, the European Commission implemented 'a global management strategy for the whole western Mediterranean' although each country still manages its own waters and governs in which areas or seasons bottom fishing is restricted (Vigo et al., 2023, p. 2).

Bottom trawling has been active throughout the Mediterranean for more than 80 years, and most fishery stocks are being captured at their maximum yield or above, putting them at unsustainable levels (Vigo et al., 2023; FAO, 2022). In

---

[1] http://www.benthis.eu

addition to overfishing and the indiscriminate nature of trawl nets, bottom trawling is also a highly destructive force that flattens and damages the benthic ecosystem on and within the seabed. This repetitive action changes the morphology of the seabed, smoothing over bedforms and other bathymetric features, as well as causes the resuspension and removal of sediment through the impacts of the wires, nets, and otter doors (Vigo et al., 2023; Puig et al., 2012). Areas of intense commercial trawling, therefore, act as an anthropogenic geological force that changes the bathymetry and benthic ecology, often scraping down far enough to expose hard substrate. Large swaths of continental shelf substrate have been damaged in this way, particularly in the Straits of Sicily, the northern Aegean Sea, and areas of the Adriatic Sea (Pulcinella et al., 2023; Ferrà et al., 2020). A recent study established a no-take area and observed the recovery of the benthos, which showed that ecological recovery following the cessation of trawling occurred in a relatively short time (Vigo et al., 2023). This is positive information for resource managers and fishery biologists. However, ancient shipwrecks cannot recover from the impacts of trawling like a benthic ecosystem can; damage to these historic sites is permanent.

A recent development in maritime shipping is the introduction of the Automatic Identification System (AIS), which was developed to avoid collisions between ships, and has been required on all vessels greater than 15 m in length since 2014 for the monitoring and management of a number of maritime activities, including fishing (Ferrà et al., 2020). The implementation of AIS on fishing vessels allows for the tracking of their mobility and operations, resulting in 'consistent data to observe large deep-sea trawlers' (Pulcinella et al., 2023, p. 810). This tracking system also shows instances where there are data gaps stemming from captains switching off the system for short periods and allowing 'hidden fishing' or unobserved bottom trawling in prohibited areas (Ferrà et al., 2020). Bottom trawling is often focused on prime bathymetric areas that contain morphologic features conducive to both the habitat necessary for the targeted catch species and ease of towing gear. However, trawling is generally prohibited within 3 miles of the coast in the Mediterranean or within the 50 m isobath as well as below 1000 m (Smith et al., 2023). Illegal trawling operations may target areas within these parameters that are otherwise prohibited, and vessels operate for short periods with their AIS transponder switched off to avoid it being reported.

4.2 Nautilus *Expeditions 2008–2012*

4.2.1 Yalıkavak

Ocean Exploration Trust began operations in the coastal waters of Turkey in 2008 while E/V *Nautilus* was in drydock near Istanbul being made ready for use the following year. A small dive boat was used to tow a side-scan sonar along survey lines around the northwestern side of the Bodrum peninsula in southwest Turkey around the town of Yalıkavak. Two late Hellenistic wrecks were already known by the Institute for Nautical Archaeology, one in the Yalıkavak harbour and the other just outside, and another four were located. Two of these consisted of rock ballast piles, likely from the nearby island of Çavuş Adasi as, historically they brought rock around from the back side to build a breakwater on the eastern side and ceramics among the rock indicated these were shipwrecks and from the Byzantine period. Another wreck, lying just outside the entrance to the natural harbour, consisted almost entirely of terra cotta pipes (Brennan, 2009).

During the sonar survey of the areas west and northwest of Yalıkavak harbour, the survey team noted significant trawl scars across the submarine landscape, particularly to the northwest. A line on the nautical chart denoted a 2.5 km boundary within which trawling operations were prohibited. The sonar indicated that the visible trawl scars on the seabed disappeared consistently at this demarcation on the chart, and even showed curves to the west as fishers began veering away while they recovered their towed gear. This indicated that the trawlers in this coastal area abided local regulations (Brennan et al., 2012a).

4.2.2 Knidos

One of the first expeditions with E/V *Nautilus* in 2009 continued the surveys off Yalıkavak and continued south of Bodrum to the northern, western, and southern coastal areas around the Datça peninsula where the ancient site of Knidos lies. Twenty-six shipwrecks were located and documented in these areas between 2009 and 2012, ranging in age from Classical Greek to Ottoman (Brennan et al., 2012a, b; Brennan & Ballard, 2013). This set of shipwrecks also ranges in impacts from trawling gear. Like those off Yalıkavak to the north, trawlers off Knidos appear to adhere to the 2.5 km from shore restriction. While none of the Knidos wrecks found are within that range, we did note that, likely due to this prohibition, trawl damage increased with distance from shore (Brennan et al., 2012a, b). The bathymetry south of Knidos is a flat, gentle southward slope, allowing for easy trawling operations parallel to the peninsula in an east-west direction.

Four other wrecks were found to the east off the coast of Marmaris, one of which that was carrying Late Roman Amphora (LRA) style amphora, in particular, showed significant trawl damage (Fig. 4.1). This wreck, Marmaris B, is easily comparable with two other LRA wrecks off Knidos (A and C), which have both been damaged by trawls, but not to the extent that Marmaris B has. These wrecks illustrate just how destructive repeated strikes from bottom trawls are to ancient shipwrecks (Brennan et al., 2020).

Fig. 4.1 ROV image of Marmaris B shipwreck (Ocean Exploration Trust)

4.2.3 Sinop

Exploration in the Black Sea along the northern Turkish coast was initiated in 1999 by Robert Ballard and the Institute for Exploration to investigate the isolated sea for preserved paleoshorelines and potentially well-preserved ancient shipwrecks lying in the anoxic waters below 155 m. Through 2003, the expeditions located and identified four shipwrecks off the coast of the Sinop peninsula. Three of these were mounds of 'carrot' shaped amphora on the continental shelf dating to the early Byzantine period, and the fourth was deeper at 325 m, well into the anoxic zone, and with wooden elements, including the mast, that were perfectly preserved (Ballard et al., 2001; Ward & Ballard, 2004; Ward & Horlings, 2008; Brennan et al., 2011). This latter shipwreck proved the theory put forth in 1976 by Willard Bascom that shipwrecks could be found well preserved in the anoxic depths of the Black Sea (Bascom, 1976).

E/V *Nautilus* returned to the southern Black Sea in 2011 to continue mapping the area off Sinop to further characterise the dynamics of the suboxic zone between the oxygenated surface waters and the anoxic layer below, and to see what the shipwrecks in the area could reveal about these processes. While exploration located an additional four ancient wrecks in various states of preservation, this work also documented significant trawl damage to wrecks along the shelf, particularly those found during the expeditions in the early 2000s (Brennan et al., 2013). Sinop A, in particular, showed significant changes to the site over a decade. More artefacts were exposed, with some clearly removed from the site, and wooden elements of the wreck were dug up out of the sediment during the decade between visits.

4.2.4 *Ereğli*

The 2011 and 2012 expeditions also conducted survey in a new area of the southern Black Sea, off the coast of Ereğli, Turkey, formerly Heraclea Pontica, where a flat shelf, like that off Sinop, lies off the coast above the continental shelf where the suboxic zone transitions to the anoxic deep waters. The main objective of these expeditions was to document internal waves between the oxic and anoxic waters moving across these shelves, and the varying levels of preservation of shipwrecks from different time periods illustrated these dynamics (Brennan et al., 2013). In addition, newly discovered wrecks in this area showed evidence of trawl damage. Ereğli A, B, and C are wooden wrecks with much of the timber still preserved but they were jumbled, with timbers pulled off site, by trawling. These wrecks with wooden elements preserved exhibit trawl damage differently than the amphora-mound wrecks in the Aegean Sea.

One wreck found in 2011, Ereğli E, is a Hellenistic era vessel that primarily consists of an artefact mound with some preserved wood, and it showed clear evidence of trawl damage (Davis et al., 2018). The standard protocol for newly discovered shipwrecks during these expeditions was to conduct a photomosaic and multibeam microbathymetry survey with sensors on the *Hercules* ROV, and this was done for Ereğli E. The 2012 *Nautilus* expedition returned to the site about 11 months later and upon visual inspection, it was clear that the site had sustained heavy trawl impacts since we first documented it. A repeated sonar survey allowed a direct comparison of the site between 2011 and 2012 and showed significant material had been removed from the wreck, including specific artefacts, both ceramic and timber, that were identified in 2011 (Brennan et al., 2016). Some artefacts were traceable to other parts of the wreck, while most were entirely missing. These sort of return visits and repeated surveys are essential in documenting and quantifying bottom trawl damage to shipwrecks.

4.3 Quantification

Quantifying the amount of damage to a shipwreck site is difficult, especially wrecks in deep water that are hard to access with any frequency or consistency. This work off Turkey implemented two ways to look at trawl damage quantitatively that allow for comparison between wrecks and establish baselines for future documentation.

The first attempt was done for ten amphora wrecks off Knidos and Marmaris. The high-resolution photomosaics from surveys of the wrecks allowed for visual analysis of the sites following the expeditions. Broken and unbroken amphoras were counted, estimating when a group of sherds likely represented 'one' amphora, to develop a total number in order to derive a percentage of those that were broken on the site. This percentage reflected the level of damage from trawls and allowed for comparison between sites in these areas. This was also plotted against distance from shore and showed that trawl damage increased farther offshore from the 2.5 km boundary where trawling was prohibited (Fig. 4.2). The most heavily trawled site, Marmaris B, exhibited 62.5% of its amphoras broken, and is also the farthest shipwreck from shore among those analysed (Brennan et al., 2012b).

The comparative analysis mentioned above of the two bathymetric surveys of Ereğli E 11 months apart was the 'first detailed documentation of how an ancient shipwreck site changes morphologically due to repeated strikes by bottom trawl gear' (Brennan et al., 2016, p. 87). The calculated difference between these two surveys is shown in Fig. 4.3. White colour indicates no change, while red shows negative change, or material removed from the site, and blue is positive change where artefacts or sediment were deposited. The greatest indicator of negative change is at the wreck itself, while some areas around the wreck show positive change due to material from the wreck being moved to the surrounding area by trawls. This shows how trawl gear scrapes shipwrecks away over time because they are bathymetric highs on a flat seabed. We estimated that about 15 m^3 was removed across an area 184 m^2 (Brennan et al., 2016). This method of return visits and repeated surveys allowed for the opportunity to truly quantify change to a shipwreck site over time by bottom trawling, and more work like this is needed to illustrate the threat posed to shipwrecks.

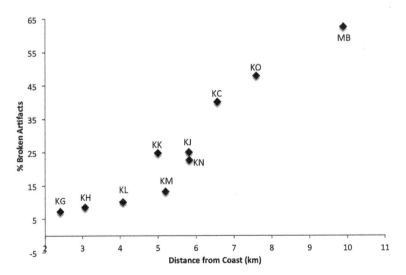

Fig. 4.2 Graph showing extent of trawl damage by distance from shore in terms of % broken amphoras. (From Brennan et al. 2012b)

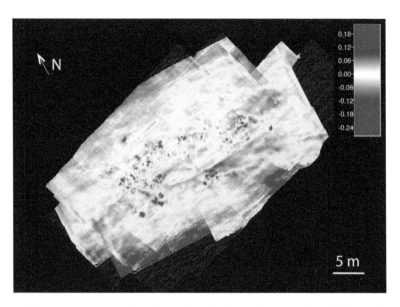

Fig. 4.3 Two sonar surveys of Eregli E conducted 11 months apart and the difference between them illustrating the material moved by trawling. (From Brennan et al., 2016).

4.4 Updated Work

While no further *Nautilus* expeditions have been conducted in Turkish waters, preventing the opportunity to revisit additional deep-water shipwrecks located between 2009 and 2012, some additional work has been undertaken since by other researchers. Exploration and documentation of ancient shipwrecks has continued in Turkey, along the eastern Aegean and northern Mediterranean coasts (Kızıldağ, 2022; Özdaş et al., 2022) as well as along the Bulgarian coast of the Black Sea (Pacheco-Ruiz et al., 2019, Prahov et al., 2021). While none of these studies looked at trawl damage directly, they made some observations to that effect.

Numerous remote sensing and photomosaic surveys were conducted by the Shipwreck Inventory Project of Turkey (SHIPT), including off the Aegean coast at Çanakkale and Yalıkavak (Kızıldağ, 2022). Five shipwrecks are reported from surveys in 2016 and 2018, three of which are amphora cargo wrecks of different ages; one of these was located by the Institute of Nautical Archaeology in 1990 in Yalıkavak harbour and denoted Yalıkavak II and is also among the wrecks documented during the *Nautilus* expeditions (Kızıldağ, 2022, Brennan et al., 2012b). Due to its location in the harbour, it is not damaged by trawls. Two amphora wrecks were found near Çanakkale, one Hellenistic and one Late Roman, and are generally in good condition—likely due to their location near the shore and within that exclusion zone discussed previously. A shipwreck consisting of a cargo of flat stones from the eighteenth century was located off the Gallipoli peninsula near Çanakkale. This wreck is located 2 km offshore and exhibited trawl damage, trawl scars in the sediment nearby, and fishing nets covering parts of the wreck. The last wreck reported by these recent surveys for SHIPT is a warship from the Ottoman period located off Bozcaada about 2.5 km from shore and was found to be heavily damaged by trawls, particularly one deep furrow that damaged the eastern side of the site (Kızıldağ, 2022). The study also notes that the use of side-scan sonar allows for the imaging of the seabed surrounding the shipwrecks and the trawl marks running through the sites, assisting with the interpretation of the damage (Kızıldağ, 2022, p. 340).

SHIPT also reports on a 2018 survey in the near-shore and coastal waters along the southern coast of Turkey around the island of Kekova Adası (Özdaş et al., 2022). This survey documented more than 25 sites along the near-shore area, many consisting of scattered and mixed assemblages due to the overlapping of activities and well-travelled routes as well as disturbance by waves action, as these sites are in shallow water. The authors note that disturbance of underwater sites, in both shallow water by coastal processes and deep water by anthropogenic activities such as fishing, 'presents a major challenge to preservation and study' (Özdaş et al., 2022, p. 14). Instead of fishing at these shallow areas, however, sites around Kekova Adası were often disturbed by 'casual collection before its declaration as a Specially Protected Area in 1990' (Özdaş et al., 2022, p. 5).

The Black Sea Maritime Archaeology Project (Black Sea MAP) was conducted in Bulgarian waters of the Black Sea from 2015 to 2017 and located 65 shipwrecks ranging from the fourth century BC to the nineteenth century AD in depths ranging from 40 to 2200 m. This survey employed deep-sea robotics for detailed photogrammetry of each shipwreck (Pacheco-Ruiz et al., 2019). Many of the wrecks discovered during this work were in the anoxic waters below 155 m depth and well preserved, but also out of the reach of trawling vessels, which do not trawl anoxic waters where fish do not live. The authors note a large difference between the wrecks located on the shelf and those in the anoxic waters due to disarticulation by trawling. For this project, however, most of the wrecks discovered were fortunately in the anoxic waters (Pacheco-Ruiz et al., 2019).

Analysis of the shipwrecks discovered during the Black Sea MAP project is ongoing. Bulgarian archaeologists have analysed their waters for litter and other anthropogenic debris, and this includes deep-water shipwrecks as additional impacts aside from trawling (Prahov et al., 2021). Bulgaria has yet to regulate bottom trawl fishing and the activity has impacted most of the continental shelf in Bulgarian waters down to depths of 100 m. As noted above, some of the Black Sea MAP wrecks were impacted by trawls, but the larger results of this project have yet to be published. However, policy and regulations are clearly needed for Bulgarian waters, for both ecological and archaeological reasons.

4.5 Marine Protected Areas

A colleague and I reviewed the species of fish visible on the numerous wrecks located during *Nautilus* expeditions in Turkish coastal waters through the ROV imagery. This led to a discussion about how shipwrecks act as hard substrate and artificial reefs and habitat for benthic ecosystems and fish to congregate at, particularly juveniles. If shipwreck sites – or in the case of areas like Sinop or Knidos, certain areas containing multiple shipwrecks – were protected in established Marine Protected Areas (MPAs), juvenile fish populations could thrive and 'spillover' into areas where fishing remained allowed, thereby both protecting the shipwreck sites and helping to increase the fishery (Krumholz & Brennan, 2015). This is a common topic in fishery sustainability research, but here applied to ancient shipwrecks. Recent research of seabed recovery from trawling continues to put forth this argument: 'The establishment of Marine Protected Areas, such as legally recognized no-take reserves where fishing activity is prohibited, could be a useful management measure... the benefits obtained from MPAs could also be observed in adjacent areas, as a result of the spillover of adults and juveniles from the protected area' (Vigo et al., 2023, p. 2).

The development of modern satellite technology now offers new options for resource management in remote areas. All vessels now carry AIS and active trawlers can be monitored (Pulcinella et al., 2023). Such use can provide the resources necessary so deep-water shipwrecks are no longer out of sight and out of mind. An option for resource managers and regulatory agencies could be to set up a type of EZPass (used for tolls on the highways in the United States) implementing trawlers' AIS. Geofences can be set up so when a trawler crosses into certain protected areas, such as areas of high numbers of shipwrecks like that off Knidos, a fine is deducted from their account. A fine can also be deducted if the AIS is turned off, indicating hidden fishing in illegal areas (Ferrà et al., 2020). This new technology can allow for real-time monitoring and protection of shipwreck sites, assuming they are first located and documented.

4.6 Conclusions

The shipwrecks located in Turkish waters during the *Nautilus* expeditions have not been revisited in the last decade, but nothing has changed in terms of trawling regulations for these waters, so it is likely that these wrecks are continuing to be dismantled by active trawling operations. The European Union and UNESCO are making headway in regulating trawling for environmental reasons, and cultural resources could be added to the discussion moving forward. European Member States are obligated to ensure that their fisheries are sustainable and that their waters are in Good Environmental Status (GES) (Smith et al., 2023). Shipwrecks can be added to this, if more support is provided to ocean exploration for documentation and return visits to these sites and repeat surveys to document damage.

Implementing modern satellite technology and improving upon AIS systems already in place can provide management options that would only require infrastructure to be developed, which regulatory agencies could do. The need, therefore, is for expanded deep-water exploration and documentation of shipwrecks worldwide that are within the depths that can be trawled. We cannot protect a resource we do not know about. Pulcinella et al. (2023) show that the Turkish fishers' exploitation off Knidos is not intense. Ancient shipwrecks that sank in more heavily trawled areas, such as the Straits of Sicily, may have been trawled to the point that they no longer exist, but those in Turkish waters may yet still be preserved.

References

Ballard, R. D., Hiebert, F. T., Coleman, D. F., Ward, C., Smith, J. S., Willis, K., Foley, B., Croff, K., Major, C., & Torre, F. (2001). Deepwater archaeology of the Black Sea: The 2000 season at Sinop, Turkey. *American Journal of Archaeology, 105*, 607–623.

Bascom, W. (1976). *Deep water, ancient ships: The treasure vault of the Mediterranean*. David & Charles.

Brennan, M. L. (2009). Ancient shipwreck survey and the modern submarine landscape off Yalikavak, Turkey. *Marine Technology Society Journal, 43*(1), 47–49.

Brennan, M. L., & Ballard, R. D. (2013). Deep-water ancient shipwrecks of the Mediterranean, Aegean, and Black Seas: 1988–2012. *Oceanography, 26*(Supplement, 1), 22–25.

Brennan, M. L., Ballard, R. D., Croff Bell, K. L., & Piechota, D. (2011). Archaeological oceanography and environmental characterization of shipwrecks in the Black Sea. In I. Buynevich, V. Yanko-Hombach, A. Gilbert, & R. E. Martin (Eds.), *Geology and Geoarchaeology of the Black Sea region: Beyond the flood hypothesis* (Geological Society of America special paper) (Vol. 473, pp. 179–188). https://doi.org/10.1130/2011.2473(11)

Brennan, M. L., Ballard, R. D., Duman, M., Tuzun, S., Inglis, G., & Turanli, T. (2012a). Continued documentation of the coastal landscape off the Datca peninsula, Turkey. *Oceanography, 25*(Supplement, 1), 28–29.

Brennan, M. L., Ballard, R. D., Roman, C., Croff, K. L., Bell, B. B., Coleman, D. F., Inglis, G., Koyagasioglu, O., & Turanli, T. (2012b). Evaluation of the modern submarine landscape off southwestern Turkey through the documentation of ancient shipwreck sites. *Continental Shelf Research, 43*, 55–70.

Brennan, M. L., Davis, D., Roman, C., Buynevich, I., Catsambis, A., Kofahl, M., Derya Ürkmez, J., Vaughn, I., Merrigan, M., & Duman, M. (2013). Ocean dynamics and anthropogenic impacts along the southern Black Sea shelf examined by the preservation of pre-modern shipwrecks. *Continental Shelf Research, 53*, 89–101.

Brennan, M. L., Davis, D., Ballard, R. D., Trembanis, A. C., Ian Vaughn, J., Krumholz, J. S., Delgado, J. P., Roman, C. N., Smart, C., Bell, K. L. C., Duman, M., & DuVal, C. (2016). Quantification of bottom trawl fishing damage to ancient shipwreck sites. *Marine Geology, 371*, 82–88.

Brennan, M. L., Davis, D., Opaiţ, A., & Stay, M. (2020). Deep-water shipwrecks as a microcosm of Late Roman maritime exchange in the Eastern Mediterranean. *Journal of Roman Archaeology, 33*, 291–330.

Davis, D., Brennan, M. L., Opaiţ, A., & Beatrice, J. S. (2018). The Eregli E shipwreck: A Hellenistic merchant ship in the Black Sea. *International Journal of Nautical Archaeology, 41*, 57–80.

Eigaard, O. R., Bastardie, F., Hintzen, N. T., Buhl-Mortensen, L., Buhl-Mortensen, P., Catarino, R., Dinesen, G. E., Egekvist, J., Fock, H. O., Geitner, K., Gerritsen, H. D., Gonz'alez, M. M., Jonsson, P., Kavadas, S., Laffargue, P., Lundy, M., Gonzalez-Mirelis, G., Nielsen, J. R., Papadopoulou, N., Posen, P. E., Pulcinella, J., et al. (2017). The footprint of bottom trawling in European waters: Distribution, intensity, and seabed integrity. *ICES Journal of Marine Science, 74*, 847–865.

FAO. (2022). *The state of world fisheries and aquaculture 2022. Towards blue transformation*. Rome. https://doi.org/10.4060/cc0461en

Ferrà, C., Tassetti, A. N., Armelloni, E. N., Galdelli, A., Scarcella, G., & Fabi, G. (2020). Using AIS to attempt a quantitative evaluation of unobserved trawling activity in the Mediterranean Sea. *Frontiers in Marine Science, 7*, 1–10.

Kızıldağ, N. (2022). Mapping and recording of ancient shipwrecks by using marine remote sensing techniques: Case studies from Turkish coasts. *Marine Science and Technology Bulletin, 11*(3), 331–342. https://doi.org/10.33714/masteb.1144180

Krumholz, J. S., & Brennan, M. L. (2015). Fishing for common ground: Investigations of the impact of trawling on ancient shipwreck sites uncovers a potential for management synergy. *Marine Policy, 61*, 127–133.

Özdaş, H., Leidwanger, J., Gross, J., & Kızıldağ, N. (2022). Toward systematic underwater survey of Mediterranean maritime activity along the southern Turkish coast. *Journal of Field Archaeology*. https://doi.org/10.1080/00934690.2022.2076189

Pacheco-Ruiz, R., Adams, J., Pedrotti, F., Grant, M., Holmlund, J., & Bailey, C. (2019). Deep sea archaeological survey in the Black Sea – Robotic documentation of 2,500 years of human seafaring. *Deep Sea Research Part I: Oceanographic Research Papers, 152*, 103087.

Pinello, D., Gee, J., Accadia, P., Sabatella, E. C., Vitale, S., Polymeros, K., & Fiorentino, F. (2018). Efficiency of shallow and deep-water trawling in the Mediterranean and its implications for discard reduction. *Scientia Marina, 82*, 97–106.

Puig, P., Canals, M., Company, J. B., Martín, J., Amblas, D., Lastras, G., Palanques, A., & Calafat, A. M. (2012). Ploughing the deep sea floor. *Nature, 489*, 286–289.

Pulcinella, J., Armelloni, E. N., Ferrà, C., Scarcella, G., & Tassetti, A. N. (2023). Deepwater red shrimp fishery in the eastern-central Mediterranean Sea: AIS-observed monthly fishing effort and frequency over 4 years. *Earth System Science Data, 15*, 809–820.

Prahov, N., Prodanov, B., Dimitrov, K., & Velkovsky, K. (2021). The nagative impact of human activities on underwater cultural heritage: Case studies from the Bulgarian Black Sea Littoral. *SGEM International Multidisciplinary Scientific GeoConference, STEF92 Technology*. https://doi.org/10.5593/sgem2021/2.1/s11.89

Smith, C. J., Papadopoulou, N. K., Maina, I., Kavadas, S., van Denderen, P. D., Katsiaras, N., Reizopoulou, S., Karakassis, I., Tselepides, A., & Tsikopoulou, I. (2023). Relating benthic sensitivity and status to spatial distribution and intensity of trawling in the Eastern Mediterranean. *Ecological Indicators, 150*, 110286.

Vigo, M., Navarro, J., Aguzzi, J., Bahamón, N., Garcia, J. A., Rotllant, G., Recasens, L., & Company, J. B. (2023). ROV-based monitoring of passive ecological recovery in a deep-sea no-take fishery reserve. *Science of the Total Environment, 883*, 163339.

Ward, C., & Ballard, R. (2004). Black Sea shipwreck survey, 2000. *International Journal of Nautical Archaeology, 33*, 2–13.

Ward, C., & Horlings, R. (2008). The remote exploration and archaeological survey of four Byzantine ships in the Black Sea. In R. D. Ballard (Ed.), *Archaeological oceanography* (pp. 148–173). Princeton University Press.

Open Access This chapter is licensed under the terms of the Creative Commons Attribution 4.0 International License (http://creativecommons.org/licenses/by/4.0/), which permits use, sharing, adaptation, distribution and reproduction in any medium or format, as long as you give appropriate credit to the original author(s) and the source, provide a link to the Creative Commons license and indicate if changes were made.

The images or other third party material in this chapter are included in the chapter's Creative Commons license, unless indicated otherwise in a credit line to the material. If material is not included in the chapter's Creative Commons license and your intended use is not permitted by statutory regulation or exceeds the permitted use, you will need to obtain permission directly from the copyright holder.

Chapter 5
Wreck Sites as Systems Disrupted by Trawling

Jan Majcher, Rory Quinn, Gert Normann Andersen, and David Gregory

Abstract This chapter examines the effects of bottom trawling on shipwreck sites, conceptualising them as process-response systems that achieve a quasi-equilibrium state over time. Disruptions to this state by bottom-contact fishing gear are analysed through examples from recent geophysical surveys in the Irish, Baltic, and North Seas. The study highlights the capabilities and limitations of modern geophysical methods in detecting changes at underwater archaeological sites caused by bottom trawling. Specifically, it addresses the challenges of identifying evidence of disturbance on dynamic seabeds and suggests that detailed analysis of wreck distribution might provide indirect proxies of structural damage due to trawling activities. Furthermore, it emphasises the potential of these disturbances to mobilise hazardous materials, such as unexploded ordnance and fuel from modern shipwrecks, posing an added environmental risk. Acknowledging existing knowledge gaps in the understanding of trawling impacts on underwater cultural heritage and the marine environment, the authors call for more case study research.

5.1 Introduction

Shipwreck sites can be considered as systems reaching a quasi-equilibrium state after the initial wrecking event (Astley, 2016; Majcher et al., 2021; Quinn, 2006; Quinn & Boland, 2010). This balanced state is fluid and is both susceptible to and dependent on external and internal influences. Any disturbances might lead to a new equilibrium state or maintain the existing one, contingent on the site's overall resilience to external forces (Quinn & Boland, 2010).

J. Majcher (✉) · D. Gregory
The National Museum of Denmark, Department of Research, Collections and Conservation, Section for Environmental Archaeology and Materials Science, Kongens Lyngby, Denmark

R. Quinn
School of Geography and Environmental Sciences, Ulster University, Coleraine, UK

G. N. Andersen
Sea War Museum Jutland, Thyborøn, Denmark

© The Author(s) 2024
C. Jarvis (ed.), *Threats to Our Ocean Heritage: Bottom Trawling*,
SpringerBriefs in Underwater Archaeology, https://doi.org/10.1007/978-3-031-57953-0_5

The internal factors can be understood as those variables which are determined before and during the ship's deposition on the seabed: its form, design, materials, function, operating conditions and the nature of the sinking (Gibbs, 2006; Gregory et al., 2012; Muckelroy, 1976). After the dynamic sinking event, the wreck begins interacting with its surrounding environment, settling onto the seabed. This is when the local ocean environment and human impacts—namely, the external factors such as bottom trawling—come into play.

The influence of the local environment depends on local and regional oceanographic and geological factors. If the archaeological material is exposed on highly energetic seabeds—affected by waves, strong currents, storms and induced sediment mobility—it will most likely undergo continuous alteration (Astley, 2016; Majcher et al., 2021; Quinn & Boland, 2010; Stieglitz & Waterson, 2013) (Fig. 5.1a). Such conditions can lead to site burial/exposure events, accelerated structural collapse from hydrodynamic forcing, and changes in the distribution of forces acting on the vessel due to sediment transport, such as mobile bedforms and seabed scour (Quinn, 2006). Conversely, the preservation of sites located in less dynamic ocean environments, with limited influence from hydrodynamic forcing, is primarily affected by processes like chemical corrosion and biological encrustation, and their associated parameters like oxygen concentration, temperature and salinity (Gregory, 2020) (Fig. 5.1b). No conditions exist in which a site deposited underwater is preserved indefinitely (Fig. 5.1c). Natural forces always induce positive entropy and a disequilibrium trend over time, even when their magnitudes are minimal.

The majority of natural processes, with the exception of heavy storms, exert a relatively weak long-term impact, especially when compared to often abrupt human impacts. Activities such as commercial fishing, illegal salvage, dredging, and seabed engineering are potentially capable of causing impulses of intense disruption to the archaeological record of an individual wreck site (Brennan et al., 2016; Quinn & Boland, 2010). Such disturbances are often challenging to identify and predict, and once detected, it is difficult to establish whether the causative force was natural or anthropogenic (Brennan, 2016), or a combination of both. Additionally, in cases where the disturbance is anthropogenic, pinpointing the exact type of human activity that caused the visible disruption can be difficult.

As outlined in other chapters of this book, commercial fishing—particularly bottom trawling—can exert a devastating impact on Underwater Cultural Heritage (UCH). There needs to be additional case studies to understand the magnitude and scale of these impacts. Such studies will pave the way for novel strategies to address the issue, encompassing both monitoring and prevention measures. This chapter delves into the potential consequences of bottom trawling on shipwreck sites, viewing them as systems at various stages of equilibrium corresponding to their surrounding environment. The discussion draws on examples from recent geophysical surveys carried out in the Irish, Baltic and North Seas. It examines the potential of contemporary geophysical techniques in tracking changes at underwater sites induced by bottom trawling. Through this discourse, the authors identify knowledge gaps related to the impact of bottom-contact fishing on UCH, as well as the challenges associated with addressing them.

5 Wreck Sites as Systems Disrupted by Trawling

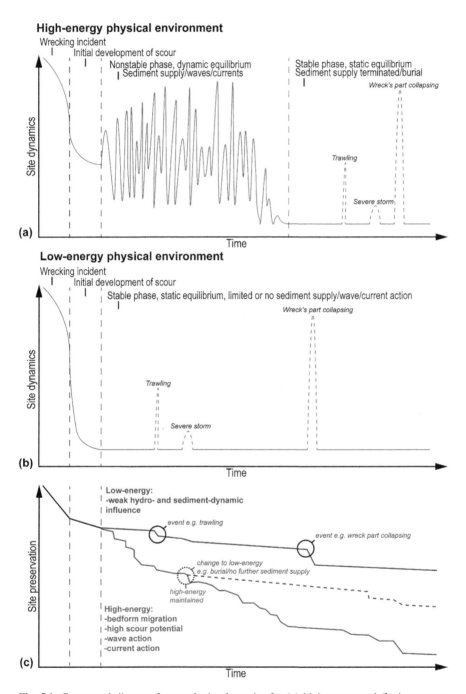

Fig. 5.1 Conceptual diagram for wreck site dynamics for (**a**) high-energy and (**b**) low energy environments, along with (**c**) corresponding estimates for site preservation in the two categories. Rapid changes in site dynamics due to either continuous sediment supply or singular events, result in accelerated degradation and lower site preservation. In contrast, periods of stability favour preservation. However, preservation is always steadily declining due to corrosion and biological action. (Adapted from Majcher et al., 2021)

5.2 Bottom Trawling as an Anthropogenic Trigger Disrupting Wreck Sites

When a wreck system experiences disruption due to external forcing, the site either absorbs the disruption, maintaining its current state, or undergoes change, reaching a new equilibrium state (Quinn & Boland, 2010). The outcome depends on the resilience of the site, which is determined both by its pre-depositional parameters (i.e., materials used to construct the hull, nature of the wrecking incident etc.) and post-depositional processes it undergoes. The composition of the seabed plays a significant role, either enabling or preventing further disruption. While the impact of bottom contact fishing has not been explored experimentally or through case studies within the 'open system' framework, potential scenarios can be inferred from known examples of natural and other anthropogenic forces disrupting sites.

The first relevant example, published by Quinn and Boland (2010), pertains to a site situated in a highly dynamic riverine environment. The *Drogheda Boat* site, near the River Boyne's outlet to the Irish Sea was discovered during dredging operations in 2006, and subsequently monitored for impacts associated with dredging. Even though much of the seabed surrounding the wreck mound had been excavated, the site itself remained undisturbed by the dredging operators. Successive geophysical surveys showed that the wreck mound was eroding rapidly, revealing artefactual material. This observation necessitated the redeposition of some sediment that had previously been dredged from around the site.

Quinn and Boland (2010) outlined how rapid erosion of the wreck mound was triggered by changes in hydrodynamic patterns associated with the dredging. They further framed this example within the 'open system' narrative, theorising that because of the external trigger (i.e., the capital dredging operation), the site exceeded its resilience threshold, transitioning to a new system state characterised by elevated erosion potential and consequent material loss. Bottom trawling possesses a similar capacity to push sites beyond their resilience thresholds, promoting erosion and/or site reorganisation.

Trawling is potentially capable of mechanically displacing or causing collapse of individual wreck site elements, leading to reorganisation of the site. The extent of this depends on the site's resilience and capacity to absorb such impacts. For instance, if the site is predisposed to erosion due to its sedimentary setting, like the *Drogheda Boat* site (Quinn & Boland, 2010), this will negatively affect the site's resilience. Specifically, displaced elements, particularly larger objects, would act as new obstacles to water flow. This alters the hydrodynamic regime at the site, effectively making these objects nuclei for further seabed scour (Quinn, 2006). Such dynamics might lead to the exposure or burial of other sections of the wreck, even if they were not originally in contact with the trawling gear. This is analogous to the dredging activity observed at the *Drogheda Boat* site. While the primary focus of dredging was directed away from the site, it inadvertently caused progressive exposure. As such, when trawling impacts a site, it is safe to assume that the resultant disturbance could extend beyond the initial point of interaction.

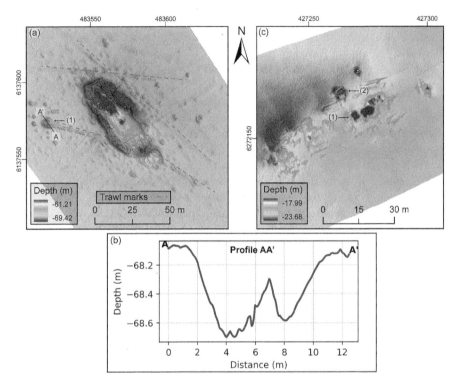

Fig. 5.2 Multibeam echosounder data-derived, hillshaded digital elevation models of (**a**) a steamship wreck, possibly *SS Edgar* in the Baltic Sea with (**b**) a bathymetric profile denoted as AA'. (**c**) hillshaded digital elevation model of an unknown wreck in the Danish North Sea. Numbers shown in on the figures correspond to structural elements discussed in the text. (Data courtesy of JD-Contractor A/S)

Convincing evidence of direct trawl damage to the wreck of a steamship (likely SS *Edgar*, lost in 1894) in the Baltic Sea is shown in Fig. 5.2a. The site is intersected by multiple interpreted trawl marks, and numerous displaced sonar contacts are scattered around the wreck. One particularly prominent example, marked as (1) in Fig. 5.2a, displays a structural element that measures roughly 4 × 1.5 × 0.3 m in terms of length, width, and height, respectively. The object is located 32 m from the wreck's hull and lies at the end of a linear trawl mark which intersects the stern section of the vessel. It is probable that this object was dislodged from the wreck and represents one of its original structural components. A scour pit measuring 10 m across and 0.5 m deep is developed around the object (Fig. 5.2b). Although, in this instance, the scour pit does not directly threaten the integrity of the hull, it shows how the site's system adapted to the redistributed wreck components on the seabed.

Another example of a site most likely impacted by trawling (Fig. 5.2b) is represented by a significantly deteriorated, unidentified wreck in the North Sea at a depth of 20 m. While a large portion of the wreck's hull has decayed, several heavily-engineered structural elements remain on the seabed, likely remnants of the ship's steam engine and machinery. Two boilers are imaged, one rests within the deteriorated hull ((1) in Fig. 5.2b) and the other is located externally, offset to the north of the vessel ((2) in Fig. 5.2b).

Multiple explanations for the second boiler's position outside of the main wreck structure are conceivable. One possibility is a salvage operation that resulted in the boiler's relocation. Another scenario is that the boiler slid out during the wrecking event. A third possibility is displacement due to trawling. While the first scenario remains plausible, the second appears less likely. This is because the ship's hull maintains a coherent shape, suggesting it did not break up or undergo a violent wrecking event to displace the boiler from its original position. Even though no prominent trawl marks are evident in the multibeam data, the third possibility is credible, depending on the specifications of the trawl gear (Brennan, 2016). Open-source Vessel Monitoring System (VMS) data from EMODnet (2023) demonstrate that beam trawls, bottom otter trawls, and bottom seines are operational in the area. Given that a significant portion of the ship's structure has decayed, the boilers, now largely exposed in the water column, pose clear obstructions to fishing gear.

As evidenced in many surveys (e.g., Fig. 5.2b) boilers are among the most resilient structural elements on shipwreck sites. This may be attributed to the fact that they were built to withstand high operating pressures and temperatures, with thick and strong iron or steel structures. Additionally, due to their location within a vessel, boilers are also sheltered from the physical forces affecting the wreck's hull externally, hence being protected by its structure until it gradually deteriorates or collapses. These factors possibly contribute to their long-term durability in the marine environment compared to the other parts of the ship.

Nevertheless, the displacement of the boiler recorded in the multibeam data (Fig. 5.2b) could have implications for its further preservation. Just as the structural element pulled out of the shipwreck in Fig. 5.2a, the displacement and exposure of the boiler to bottom currents led to the formation of a small scour pit around it. The boiler that remained within the hull is not affected by seabed scour, possibly because it is shielded by the hull remains. Therefore, the displaced boiler is more exposed to the physical environment, after losing the protective function of the wreck's structure. While both boilers appear to be in good condition presently, this example underscores the destructive potential that trawling or other human activities capable of causing displacement might have, especially when juxtaposed with natural forces. The future preservation potential of the boiler removed from the wreck's hull may be compromised due to that displacement event.

Given the examples mentioned here, it can be inferred that the detrimental impacts of bottom-contact fishing are not confined to ancient sites (Brennan et al., 2016). They also extend to larger, relatively more coherent modern wrecks, like the steamship shown in Fig 5.2a. In essence, a trawling event acts as a 'scrambling device' (Muckelroy, 1976), initially triggering reorganisation of a wreck site. This site may then undergo further changes, depending on its capacity to absorb the disruption. Potential outcomes might involve seabed erosion leading to heightened exposure of relocated structural components to external forces, and result in the acceleration of corrosion process.

5.3 Challenges in Detecting Trawling Damage at Wreck Sites Using Geophysical Methods

When viewing an underwater wreck site through the lens of an open system—one that begins with a set of pre-depositional conditions and can adapt or change in response to various influences, both natural and anthropogenic—it becomes imperative to identify the primary drivers of these changes. Recent attempts at quantifying the rates of geomorphic change (e.g. scour, bedform migration) and associated hydrodynamic triggers were discussed by Majcher et al. (2021) using this 'open system' concept, inspired by earlier investigations (Astley, 2016; Quinn, 2006; Quinn & Boland, 2010). Geomorphic change can be effectively recorded using high-resolution geophysical data in a time-lapse sense. Furthermore, if the data collected over a shipwreck are of sufficient spatial resolution, an initial assessment of the wreck's structural changes can be conducted. However, as discussed in this chapter, determining whether the observed changes were caused by natural or anthropogenic forces presents a complex challenge.

Recent advancements in geophysical techniques have afforded researchers unprecedented insights into the dynamics of underwater shipwreck sites. For example a recent study captured centimetric-scale alterations in the seabed surrounding several metal-hulled shipwrecks in the Irish Sea (Majcher et al., 2021). The investigation determined that some of the sites, located on sandy seabeds with large active tidal bedforms, undergo constant change with high volumes of sediment transported through the sites. For example, between 2015 and 2019, the SS *WM Barkley* site experienced a remarkable elevation change of 4.9 m in one area, partly exposing the starboard side of the vessel (Fig. 5.3a). To investigate these rapid changes, Computational Fluid Dynamic (CFD) simulations were used to model tidal current flows which supported the understanding of the hydrodynamic patterns at the site and explained the pervasive bedform movement and seabed scour.

In addition to investigating geomorphic adjustment at these sites, an attempt was made to detect structural changes at these wrecks through point-cloud comparisons of high density multibeam data. Although the subjective nature of the manual point cloud cleaning process meant that detecting minor centimetric changes was challenging due to the noise associated with wreck data, the method was sensitive enough to identify substantial structural displacements. For example, between 2015 and 2019, the SS *WM Barkley* wreck suffered the detachment of a considerable part of its portside gunwale near the stern (Fig. 5.3a).

The environmental dynamics and the nature of the SS *WM Barkley* shipwreck's degradation present a compelling argument for the involvement of anthropogenic external forces, such as trawling, as catalysts for the observed changes. The wreck is tilted towards its starboard side, and if the gunwale had collapsed due to corrosion only, it would be expected to have moved gravitationally towards the deck. However, it moved in the opposite direction, towards the seabed. The possibility of an external force, like an anchor or trawl door pull, causing this movement becomes more likely given the direction of detachment.

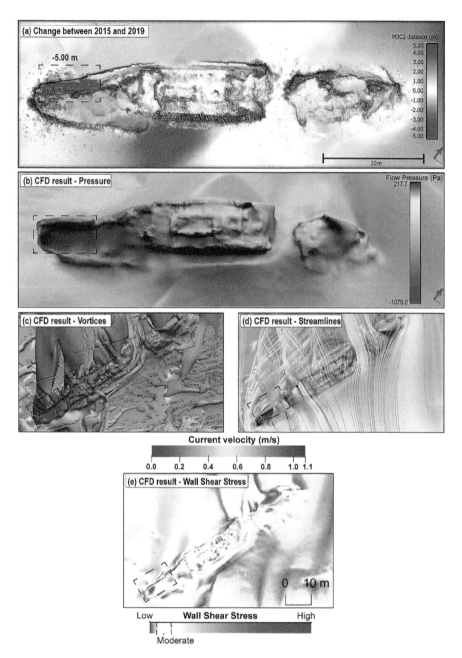

Fig. 5.3 (**a**) Differences between multibeam echosounder point clouds obtained for the SS *WM Barkley* site in 2015 and 2019. (**b**) CFD-simulated NNW-oriented tidal current flow pressure exerted on the modelled seabed at the same site, (**c**) vorticial patterns delineated by the simulations, (**d**) NNW-oriented current flow streamlines, and (**e**) wall shear stress. Dashed boxes show the stern and portside gunwale of the ship discussed in the text. A detailed data and methodology description is provided in the original publication of Majcher et al., 2022

The distant location of SS *WM Barkley* from significant ports (30 km from the port of Dublin) reduces the likelihood of anchor drags. Instead, it emphasises the potential impact of trawling activities in the area. As evidenced by the multibeam data, SS *WM Barkley* is located in a very dynamic environment, dominated by mobile sediment and strong bi-directional tidal currents. Pervasive sandwaves are capable of quickly masking any evidence of trawl scars. Although it has been shown that trawl marks can remain visible for 2–7 months in dynamic environments like inter-tidal basins (Brylinsky et al., 1994), significant geomorphic change and sand-wave migration were detected even at a weekly time intervals at the SS *WM Barkley* site (Majcher et al., 2021), potentially eradicating any trawling evidence.

Therefore, to understand possible causes of the gunwale displacement, CFD-modelled variables (Fig. 5.3b–e) were compared against structural changes (Majcher et al., 2022). A few observations were made: (1) the stern section is generally subject to relatively high flow -exerted pressure under the NNW, flow, tidal current (Fig. 5.3b); (2) vortices are created over the gunwale (Fig. 5.3c) with streamline contraction and increase in the flow speed (Fig. 5.3d), and (Fig. 5.3) the collapsed gunwale was subject to high shear-stress exerted under the NNW flow conditions (Fig 5.3e). On the other hand, the modelled tidal current flowing in the opposite, ebb, SSE direction (tidal currents are bi-directional in the area) exerted lower shear stress on the same gunwale. Although these observations support the idea that the deterioration of the gunwale could be due to natural causes, specifically the NNW tidal current-induced push towards the wreck's portside, the influence of fishing cannot be ruled out. It remains probable that the tidal flows caused gradual wear which enabled a passing trawl door or net to eventually detach the gunwale.

A contrasting Irish Sea example, where bottom trawling is assumed to directly impact a site (similar to the Baltic site of SS *Edgar* discussed above), is the FV *St. Michan* (Majcher et al., 2021). *St. Michan,* a motor fishing trawler, was lost in 1918 at a depth of 70 m, close to the Western Irish Sea Mudbelt, an area heavily trawled for Dublin Bay prawn (*Nephrops Norvegicus*) (Coughlan et al., 2015). The site was surveyed with high-resolution multibeam echosounder data in 2015 and 2019, and recent trawl marks directly intersecting the wreck were detected in the 2019 data (Fig. 5.4). Although no substantial damage was observed to the wreck's structure in the multibeam data, it cannot be ruled out, as the vessel is relatively small (30 m long) and lying exposed in a scour pit.

The examples provided here show that the degree of ambiguity in determining the influence of bottom trawling on UCH using non-invasive geophysical methods depends on many things, including local seabed conditions and the timing of surveys relative to trawling activities. It is nearly impossible to directly detect evidence of trawling on dynamic seabeds, due to high rates of sediment transport masking trawl scars. Detailed examination of wreck distribution on the seabed, aimed at finding displaced structural elements and assessment of sediment- and hydro-dynamic conditions may provide some clues about whether structural damage may be attributed to anthropogenic activities such as bottom trawling. Additionally, analysis of fishing activity data, for example based on AIS or VMS tracking (EMODNet, 2023)

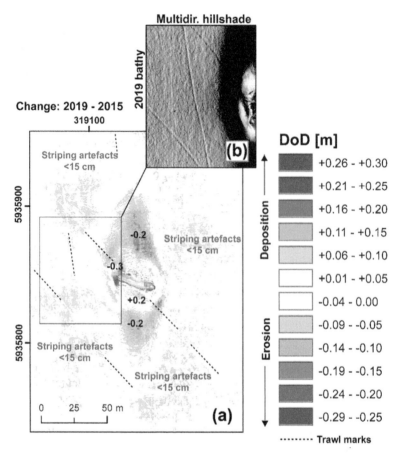

Fig. 5.4 (**a**) Digital Elevation Model of Differences (DoD) for the FV *St. Michan* site, showing bathymetric changes between years 2015 and 2019, and (**b**) inset map showing a hillshaded elevation model created using the data collected in 2019. New trawl marks were recorded in the 2019 data. (Adapted from Majcher et al., 2021)

can provide information about trawling intensity in the area. Following this, more detailed examination of wrecks could be carried out by visual inspection with a remotely operated vehicle or by diving and photogrammetric techniques to detect primary damage or other proxies of trawling e.g. presence of ghost nets (Pedersen et al., 2022). Conversely, in more static seabed environments, with minimal sediment transport, trawling can easily be evidenced and monitored using high resolution bathymetry, backscatter or side-scan sonar data (Brennan et al., 2016; Gournia et al., 2019; Majcher et al., 2021).

5.4 Conclusion and Further Research

The preservation of historic wrecks on the seabed is controlled by a complex set of changing environmental and anthropogenic variables. Considering underwater sites as process-response systems at some equilibrium state with their environment, enables an in-depth analysis of a degree of influence of individual factors and their potential to cause disruption affecting long-term preservation. Bottom-contact fishing can be considered one element of the system. Such an abrupt, high-impact trigger can lead to significant disruption of a site, permanently displacing structural elements, which in turn may lead to wholescale changes at the site in terms of dominant processes, such as a new scour regime.

However, determining the exact combination of factors responsible for particular damage recorded at a site is challenging. High-resolution geophysical methods do provide some proxies of trawling under the right seabed conditions; if the seabed is relatively static with no significant sediment transport, trawl marks will be recorded. Conversely, identification of trawl scars at highly dynamic sites with mobile sediments is often impossible. In the latter case, visual inspection may provide information about the presence of anthropogenic interference, such as ghost nets. Additional information can be sourced from open-source portals providing fishing intensity data or delineating common fishing areas. Case studies conducted in the Danish Baltic Sea (Pedersen et al., 2022) and the British North Sea (Revill & Dunlin, 2003), found the presence of ghost nets on 11 out of 18 investigated wrecks located in known fishing areas. This shows high likelihood (>50%) of direct interaction between fishing gear and a shipwreck if it is indeed located within a fishing zone.

In order to reduce the gap in knowledge pertaining to the influence of anthropogenic factors like fishing on shipwrecks, extending the sample of case studies is necessary. A quantitative statistical assessment could shed light on the scale of the problem. A new study conducted in terms of the ENDURE project (www.endureerc.com), described in Chap. 6 and Gregory et al. (2024), presents a possible further research direction by integrating geophysical datasets (high-resolution multibeam echosounder scans) with various metocean, oceanographic and human activity information sourced online. Another novel approach considers a shipwreck site as a complex adaptive system, which can be studied using Agent Based Modelling (ABM) (Vega-Sánchez & Herrera, 2022). Various pre-depositional and post-depositional factors are included in the ABM conceptual model, which could refine our understanding and determine pathways of deterioration for individual wreck sites.

Understanding anthropogenically-induced preservation or deterioration of shipwreck sites is crucial not only for their heritage value, but also from a potentially polluting perspective. Many modern wrecks, especially those from World War II, contain fuels and other hazardous materials like unexploded ordnance which are potentially detrimental to the ocean environment. Assessing their stability on the seabed is vital for environmental health (Carter et al., 2021; Szafrańska et al., 2021).

As demonstrated in this chapter, bottom-contact fishing has the potential to displace large and heavy structural elements, such as steam boilers. Consequently, it can be inferred that trawling is also capable of moving unexploded ordnance and triggering the sudden release of pollutants to the ocean environment. This potential for pollution introduces another dimension to the importance of understanding and monitoring the impact of trawling on shipwreck sites.

Acknowledgements The authors thank Rasmus Normann Andersen of JD-contractor A/S for granting us permission to use the multibeam echosounder data presented in the paper for the two North and Baltic Sea wrecks. We also extend our gratitude to Ruth Plets (Flanders Marine Institute in Belgium), Chris McGonigle (Ulster University), Fabio Sacchetti (Marine Institute Ireland), Thomas Smyth (University of Huddersfield), Mark Coughlan (University College Dublin) and Kieran Westley (Ulster University), who took part in project planning, acquiring, processing and analysing the data acquired for the mentioned Irish Sea wrecks in terms of the Marine Institute's (Ireland) ship-time applications: APP-CV15021, CV16031 and CV19027.

This work was supported by the Danish Ministry of Culture under Grant FPK.2017-0037 and the European Union (ERC, ENDURE, 101053993). Views and opinions expressed are however those of the authors only and do not necessarily reflect those of the European Union or the European Research Council Executive Agency. Neither the European Union nor the granting authority can be held responsible for them. The authors report that there are no competing interests to declare.

References

Astley, A. (2016). *The taphonomy of historic shipwreck sites* [Doctoral thesis, University of Southampton]. https://eprints.soton.ac.uk/402317/

Brennan, M. (2016). Quantifying impacts of trawling to shipwrecks. In M. E. Keith (Ed.), *Site formation processes of submerged shipwrecks* (pp. 157–179). University Press of Florida.

Brennan, M. L., Davis, D., Ballard, R. D., Trembanis, A. C., Vaughn, J. I., Krumholz, J. S., Delgado, J. P., Roman, C. N., Smart, C., Bell, K. L. C., Duman, M., & DuVal, C. (2016). Quantification of bottom trawl fishing damage to ancient shipwreck sites. *Marine Geology, 371*, 82–88. https://doi.org/10.1016/j.margeo.2015.11.001

Brylinsky, M., Gibson, J., & Gordon, D. C., Jr. (1994). Impacts of flounder trawls on the intertidal habitat and community of the Minas Basin, Bay of Fundy. *Canadian Journal of Fisheries and Aquatic Sciences, 51*(3), 650–661. https://doi.org/10.1139/f94-066

Carter, M., Goodsir, F., Cundall, P., Devlin, M., Fuller, S., Jeffery, B., Hil, G., & Talouli, A. (2021). Ticking ecological time bombs: Risk characterisation and management of oil polluting World War II shipwrecks in the Pacific Ocean. *Marine Pollution Bulletin, 164*, 112087. https://doi.org/10.1016/j.marpolbul.2021.112087

Coughlan, M., Wheeler, A. J., Dorschel, B., Lordan, C., Boer, W., van Gaever, P., de Haas, H., & Mörz, T. (2015). Record of anthropogenic impact on the Western Irish Sea mud belt. *Anthropocene, 9*, 56–69. https://doi.org/10.1016/j.ancene.2015.06.001

EMODNet. (2023). *European marine observation and data network—GeoViewer [dataset]*. https://emodnet.ec.europa.eu/geoviewer/

Gibbs, M. (2006). Cultural site formation processes in maritime archaeology: Disaster response, salvage and Muckelroy 30 years on. *International Journal of Nautical Archaeology, 35*(1), 4–19. https://doi.org/10.1111/j.1095-9270.2006.00088.x

Gournia, C., Fakiris, E., Geraga, M., Williams, D. P., & Papatheodorou, G. (2019). Automatic detection of trawl-marks in sidescan sonar images through spatial domain filtering, employing Haar-like features and morphological operations. *Geosciences, 9*(5), 214. https://doi.org/10.3390/geosciences9050214

Gregory, D. (2020). Characterizing the preservation potential of buried marine archaeological sites. *Heritage, 3*(3), 838–857. https://doi.org/10.3390/heritage3030046

Gregory, D., Jensen, P., & Strætkvern, K. (2012). Conservation and in situ preservation of wooden shipwrecks from marine environments. *Journal of Cultural Heritage, 13*(3), S139–S148. https://doi.org/10.1016/j.culher.2012.03.005

Gregory, D., Dam, M., Majcher, J., Matthiesen, H., Normann-Andersen, G., & Quinn, R. (2024). Using open-data portals, remote sensing and computational modelling to investigate historic wreck sites and their environments: 45 years on from Muckelroy. To be submitted to *International Journal of Nautical Archaeology*. https://doi.org/10.1080/10572414.2024.2320774

Majcher, J., Quinn, R., Plets, R., Coughlan, M., McGonigle, C., Sacchetti, F., & Westley, K. (2021). Spatial and temporal variability in geomorphic change at tidally influenced shipwreck sites: The use of time-lapse multibeam data for the assessment of site formation processes. *Geoarchaeology, 36*(3), 429–454. https://doi.org/10.1002/gea.21840

Majcher, J., Quinn, R., Smyth, T., Plets, R., McGonigle, C., Westley, K., Sacchetti, F., & Coughlan, M. (2022). Using difference modelling and computational fluid dynamics to investigate the evolution of complex, tidally influenced shipwreck sites. *Ocean Engineering, 246*, 110625. https://doi.org/10.1016/j.oceaneng.2022.110625

Muckelroy, K. (1976). The integration of historical and archaeological data concerning an historic wreck site: The 'Kennemerland'. *World Archaeology, 7*(3), 280–290. https://doi.org/10.1080/00438243.1976.9979641

Pedersen, E. M., Andersen, N. G., Egekvist, J., Nielsen, A., Olsen, J., Thompson, F., & Larsen, F. (2022). Ghost nets in Danish waters. *DTU Aqua*, 394–2021, 83. https://www.aqua.dtu.dk//media/institutter/aqua/publikationer/rapporter-352-400/394-2021_ghost-nets-in-danish-waters.pdf

Quinn, R. (2006). The role of scour in shipwreck site formation processes and the preservation of wreck-associated scour signatures in the sedimentary record—Evidence from seabed and sub-surface data. *Journal of Archaeological Science, 33*(10), 1419–1432. https://doi.org/10.1016/j.jas.2006.01.011

Quinn, R., & Boland, D. (2010). The role of time-lapse bathymetric surveys in assessing morphological change at shipwreck sites. *Journal of Archaeological Science, 37*(11), 2938–2946. https://doi.org/10.1016/j.jas.2010.07.005

Revill, A. S., & Dunlin, G. (2003). The fishing capacity of gillnets lost on wrecks and on open ground in UK coastal waters. *Fisheries Research, 64*(2–3), 107–113. https://doi.org/10.1016/S0165-7836(03)00209-1

Stieglitz, T. C., & Waterson, P. (2013). Impact of Cyclone Yasi on the wreck of the SS Yongala documented by comparative multibeam bathymetry analysis. *Queensland Archaeological Research, 16*, 33. https://doi.org/10.25120/qar.16.2013.222

Szafrańska, M., Gil, M., & Nowak, J. (2021). Toward monitoring and estimating the size of the HFO-contaminated seabed around a shipwreck using MBES backscatter data. *Marine Pollution Bulletin, 171*, 112747. https://doi.org/10.1016/j.marpolbul.2021.112747

Vega-Sánchez, R., & Herrera, J. M. (2022). Agent-based modelling for the study of shipwreck site formation processes: A theoretical framework and conceptual model. *F1000Research, 11*, 1525. https://doi.org/10.12688/f1000research.125089.1

Open Access This chapter is licensed under the terms of the Creative Commons Attribution 4.0 International License (http://creativecommons.org/licenses/by/4.0/), which permits use, sharing, adaptation, distribution and reproduction in any medium or format, as long as you give appropriate credit to the original author(s) and the source, provide a link to the Creative Commons license and indicate if changes were made.

The images or other third party material in this chapter are included in the chapter's Creative Commons license, unless indicated otherwise in a credit line to the material. If material is not included in the chapter's Creative Commons license and your intended use is not permitted by statutory regulation or exceeds the permitted use, you will need to obtain permission directly from the copyright holder.

Chapter 6
Multi-use and Multi-vocal Challenges of Preserving Underwater Cultural Heritage in Stellwagen Bank National Marine Sanctuary

Calvin Mires, Benjamin Haskell, and Kirstin Meyer-Kaiser

6.1 Introduction

The cultural identity of Massachusetts is intrinsically linked to the sea. For hundreds of years, ship-borne trade, migration, whaling, and fishing have tied Massachusetts to a world system of maritime commerce and culture and brought the world back to it through the area that is now Stellwagen Bank National Marine Sanctuary (SBNMS). Established in 1992, SBNMS is an 842-square mile area in the mouth of Massachusetts Bay, located 25 miles east of Boston, 3 miles southeast of Cape Ann, and 3 miles north of Provincetown, Massachusetts. Named for Lt. Henry S. Stellwagen, a nineteenth-century Navy surveyor, the sanctuary protects a productive and diverse marine ecosystem which has been the subject of centuries of fishing activity. It is also estimated that 200 shipwrecks rest on the seafloor within the sanctuary's boundaries, representing more than 400 years of maritime history, industry, and culture (Fig. 6.1).

6.2 The Problem

SBNMS is one of 15 United States marine protected areas (MPAs) under the National Marine Sanctuaries Act (NMSA) and 2 marine national monuments under the Antiquities Act within the National Marine Sanctuary System (NMSS) managed

---

C. Mires (✉) · K. Meyer-Kaiser
Woods Hole Oceanographic Institution, Woods Hole, MA, USA
e-mail: cmires@whoi.edu

B. Haskell
Stellwagen Bank National Marine Sanctuary, Massachusetts Bay, MA, USA

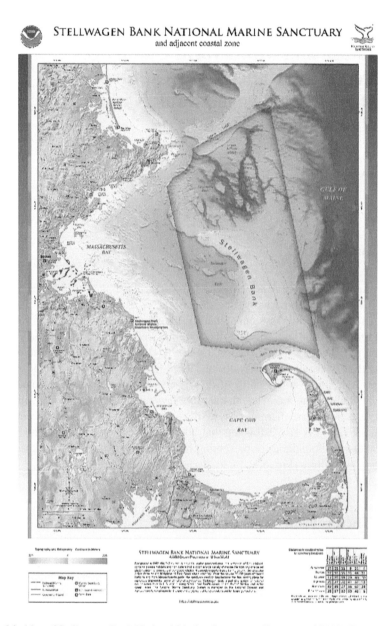

Fig. 6.1 Map of Stellwagen Bank National Marine Sanctuary. (Courtesy of SBNMS)

by the Office of National Marine Sanctuaries (ONMS), a part of the National Oceanic and Atmospheric Administration (NOAA). Due to each MPA's unique differences in settings, resources, and threats, each has a tailored management plan (Lawrence & Marx, 2011; Lawrence et al., 2015; Mires et al., 2020; Harrelson et al., 2022).

As envisioned by Congress in the NMSA, they are not 'sanctuaries' in the traditional sense of the word but are multiple-use areas with a mandate to facilitate public and private activities that are compatible with the primary objective of the MPA: resource protection. When SBNMS was designated by Congress, it carried the stipulation that recreational and commercial fishing would be allowed and continue to be controlled under existing fisheries management law. In no small part because fishing is a key component of the maritime history, identity, and economy in New England. There is strong public pressure to continue to allow these activities, and everywhere in SBNMS is currently open to fishing except for a small area (known as 'the sliver') that is closed under prior federal fishing law to commercial bottom-tending mobile and fixed gear and overlaps with a closed area in the Western Gulf of Maine. However, these activities have also proven to be destructive to Underwater Cultural Heritage (UCH) throughout the sanctuary (Meyer-Kaiser et al., 2022b; Harrelson et al., 2022; Mires & Meyer-Kaiser, 2023). In fact, fishing gear impacts were identified as the greatest threat to archaeological resources in the sanctuary (ONMS, 2020; SBNMS, 2021). Figure 6.2 visualises the complexity of the problem, showing the high degree of overlap between the shipwrecks (black dots) with the intensity of commercial fishing activity (background colour range). Therefore, SBNMS managers are left in a difficult and ironic situation: charged to protect resources under the NMSA and the National Historic Preservation Act while also unable to regulate the very activity and equipment that puts them at risk. Finding a solution is the reality, and challenge, SBNMS managers face.

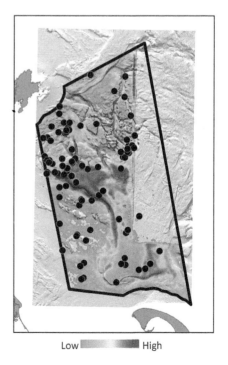

Fig. 6.2 Heatmap with shipwrecks (black dots show approximate locations) overlayed on areas of fishing activity. Red areas indicate high fishing intensity; Green, low intensity. Gray area is 'the sliver' with fishing prohibited (SBNMS, 2023a)

6.3 Underwater Cultural Heritage in SBNMS

Sanctuary historic resources, like UCH, refers to all traces of anthropogenic existence and activities with cultural, historical, or archaeological character that have been submerged for a period (UNESCO, 2001). Cultural heritage in general—and UCH specifically here—have an inherent multivocality, which means a single heritage resource in space, place, and landscapes, can be viewed simultaneously in a variety of ways (Graham et al., 2000). This multivocality is part of a dynamic, discursive process that is often filled with contention and dissonance since space, place, and landscape are constantly mutating concepts characterised by a complexity of imagery and symbolism (Pile & Keith, 1997; Osborne, 1998; Brundage, 2000; Graham et al., 2000; Seaton, 2001). Heritage is fundamental to constructs of identity because it allows an individual or group to associate itself with a particular interpretation of the past. Thus, heritage provides social benefits of value but can also underly a particular worldview (Lowenthal, 1996; Rosenzweig & Thelen, 1998; Graham et al., 2000).

Value, like heritage, involves multivocality as various agents (individual or groups) make behavioural choices based on their respective preferences and perspectives. When these preferences conflict, intense feelings arise between competing stakeholders (Graham et al., 2000) and may be heightened because of the dual nature of cultural and economic value. Cultural value is most often associated with 'non-use' values of preservation and intrinsic worth whereas economic value places a value on the use or exploitation of a resource.

Archaeology and heritage are strongly associated with each other. The contributions of archaeology to heritage are usually discussed in terms of providing cultural value, such as new information, authenticity, commemoration, symbolism, and continuity of cultural identity and sense of place. Archaeological sites encompass aspects of space, place, and landscape and have been used to reinforce or refute how heritage is perceived or valued (Trigger, 2009; Graham et al., 2000; Kristiansen, 2000; Lipe, 2002; McManamon, 2002; Jameson & Scott-Ireton, 2007).

In SBNMS, there are more than 200 hundred shipwrecks of which 50 have been located, 12 identified, and 7 of those are now listed on the National Register of Historic Places (see Table 6.1 for identified shipwrecks). They are often referred to as time capsules and while this is true, they are also rich databases that may be studied as artefacts themselves, carrier of artefacts, microcosms of maritime cultures and associated systems, sites of trauma and commemoration, or combinations of all the above (Gould, 1983; Murphy, 1983; Lawrence et al., 2015). Careful analysis of shipwrecks has great potential to alter our understanding of local, regional, national, and global communication and interactions, and provide new information about the expansion, trading patterns, and contributions from disenfranchised groups (Watson, 1983; Murphy, 1983; Flatman, 2003).

Shipwrecks are ultimately part of an environmental and ecological system that was never intended to exist. UCH attracts a range of biological organisms and serves as isolated, island-like habitats. UCH is first colonised by microorganisms, which

Table 6.1 List of Identified Shipwrecks within Stellwagen Bank National Marine Sanctuary

Ship	Vessel type	Built	Length (ft)	Lost	Reason for loss
Acme Ii	Stern trawler	1966	69	1988	Engine room flooded
Edna G[a]	Eastern-rig Dragger	1956	54	1988	Foundered
Frank A Palmer[a]	Four-masted Coal Schooner	1897	274.5	1902	Collided with *Louis B. Crary*
Heroic	Minesweeper	1941	97	1969	Fire
Joffre[a]	Two-masted Eastern-rig Dragger	1918	105.5	1947	Fire
Josephine Marie	Stern Trawler	1969	80	1992	Foundered
Lamartine[a]	Two-masted Granite Schooner	1848	79.9	1893	Foundered due to heavy seas
Louise B. Crary[a]	Five-masted Coal Schooner	1900	267	1902	Collided with *Frank A. Palmer*
North Star	Eastern-rig Dragger	1967	55	2003	Capsized
Patriot	Western-rig Dragger	1997	62	2009	Capsized
Paul Palmer[a]	Five-masted Coal Schooner	1902	276	1913	Fire
Portland[a]	Passenger Steamship	1889	290	1898	Sank in the Portland Gale of 1898

[a]Indicates vessel listed on the National Register of Historic Places (NHRP; SBNMS, 2023b)

may utilise the novel materials as substrates for metabolic reactions (Price et al., 2020; Hamdan et al., 2021). Invertebrates such as sponges and cnidarians capitalise on the hard-bottom habitats offered by UCH (Perkol-Finkel & Benayahu, 2007; Meyer-Kaiser et al., 2022a). It can take centuries for the community on UCH to match the species composition of the background community (Gravina et al., 2021), and in some cases, structural differences mean the communities never match (Perkol-Finkel et al., 2006). The net effect of UCH is usually an increase in local or regional biodiversity (Paxton et al., 2019).

The use of human-made structures as artificial reefs has been documented since the 1700s. Anthropogenic activities on and around UCH inevitably contribute to site formation over time. In SBNMS, UCH attracts anthropogenic activity primarily from the diving and fishing industries. Diving on shipwrecks ranges from simple visitation and photography to looting and salvage activities. Contact from divers, particularly unskilled individuals with poor buoyancy control, can damage UCH (Siciliano et al., 2016). The diving industry in SBNMS is not large but contains a few key players: locally-owned companies with a loyal base of skilled divers. Because SBNMS is an offshore sanctuary and relatively deep (i.e., 30 m or deeper), the market consists primarily of experienced divers who live locally rather than student divers or tourists. Experienced divers who are aware of sanctuary regulations present a low risk to the integrity of UCH.

Fishing near or on UCH in SBNMS includes trawling, scallop dredging, gillnets, recreational gear, and lobster pots (Harrelson et al., 2022). The history of fishing in

SBNMS dates back centuries before the area was designated a sanctuary and is a key facet of heritage in the region. In fact, some families have been involved in the commercial fishing industry for generations. Fishing and livelihoods from the sea are interwoven with cultural identity in Massachusetts and across New England. Today, the lobster and scallop fisheries are major economic forces in the region. Commercial fishers are also organised in strong trade groups, which have a public voice. These include the Massachusetts Lobsterman's Association, the Northeast Seafood Coalition, and the Cape Cod Commercial Fisherman's Alliance.

Finally, SBNMS has an obligation under the US National Historic Preservation Act of 1966 to identify, assess, and mitigate any harmful actions and effects of undertakings on historic properties (such as UCH) eligible for or listed on the National Register of Historic Places (NRHP). This includes consideration of public views and opinions (Section 106). It assumes responsibility for the preservation of historic resources that fall under its jurisdiction and identify other historic properties eligible for listing on NRHP (Section 110).

Since its inception, SBNMS has used a policy of non-disclosure to try and balanced its dual mandate of protecting UCH and facilitating compatible uses of sanctuary resources. The locations of all known shipwrecks were kept confidential to prevent looting by divers and intentional targeting of shipwrecks for fishing. In 2017, however, it became clear that this policy of non-disclosure was ineffective after intensive scallop fishing occurred on the northwest corner of Stellwagen Bank. Side-scan surveys were conducted to assess the damage to various sites after the season closed. Figure 6.3 illustrates damage to *North Star,* a modern fishing vessel, which was scattered across the seafloor significantly losing site articulation, integrity, and an unknown number of artefacts. In 2019–2020, a cooperative agreement between ONMS and Woods Hole Oceanographic Institution enabled research on three UCH sites in SBNMS (Mires et al., 2020). The project revealed severe fishing damage to multiple sites, including new entanglements of fishing nets that had not been observed during the previous survey in 2009. The primary risk to UCH is commercial fishing activity and gear impacts.

6.4 Impacts

The most obvious impact of fishing on UCH is the entanglement of ghost gear, which damages shipwreck structures. Repeatedly targeting the area exacerbates this problem and puts the UCH at higher risk for permanent loss (Brennan, 2016). For example, sometime between 2009 and 2019, fishing gear ripped almost 40-m^2 of the fantail-stern from the passenger steamer, *Portland,* (Fig. 6.4) which sank with an estimated 200 people onboard, during the storm which bears its name, the Portland Gale of 1898. The fantail now lies on the seafloor with remnants of a gillnet draped across it (Mires & Meyer-Kaiser, 2023).

Figure 6.5 shows another example from the Portland Gale, the shipwreck presumed to be the coal schooner *King Philip*. The side-scan image on the left shows

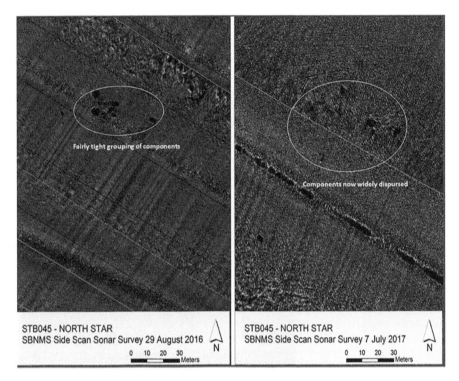

Fig. 6.3 The fishing vessel, *North Star*, in 2016 before (left) and after a scallop derby in 2017 (right) where the site has suffered wide scattering and dispersal of features due to trawling activity; dredge marks are visible in the debris field. (Image courtesy of SBNMS)

multiple dredge marks running northeast to southwest and on the right, the image clearly shows part of the starboard bow disarticulated from the wreck with dredge marks running over it. Further, lost gillnets or monofilament lines caught on shipwrecks can destroy and unintentionally entangle artefacts, snag, and break structures, and relocate artefacts.

Ghost gear continues to fish after abandonment in the ocean as fish can become entangled in ghost trawl nets (Ross et al., 2016), and two seals have drowned while caught in a ghost trawl net ensnared on the *Patriot* shipwreck in 2019 (V. Malkoski, MA DMA, pers. comm.). Fishing gear also impacts biological communities in less obvious ways. Repeated trawling along a shipwreck's hull damages or scrapes off three-dimensional sessile invertebrates, including sponges and corals that are key foundation species in the community. Areas of shipwrecks with entangled ghost gear have lower species richness and lower evenness than unaffected areas and they tend to be dominated by opportunistic species (Meyer-Kaiser et al., 2022b; Fig. 6.6). Trawling damage to a shipwreck's structure causes a change in the microhabitats available for colonisation and can lead to shifts in community composition (Mires & Meyer-Kaiser, 2023; Fig. 6.7).

It is not always clear from examination of UCH whether fishing activity was targeted on the site. Some ghost gear may represent nets that were lost off-site and

Fig. 6.4 Top: Sidescan of *Portland* in 2010 with intact stern; Middle: In 2021 with stern missing; Bottom: Photogrammetric model of *Portland* showing fishing gear draped over and around damaged stern area. (Original Images courtesy of SBNMS; Mind Technology; and Marine Imaging Technologies)

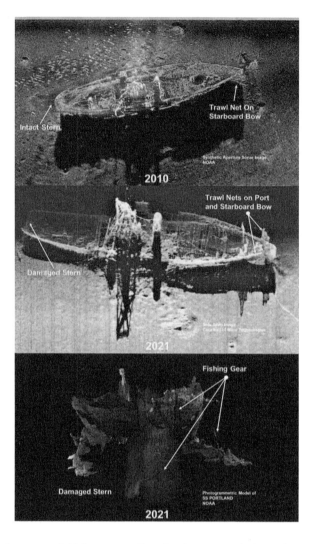

accumulated on the wreck because of tidal currents. Despite the economic incentive to target dense fish populations, shipwrecks impact the safety of captains and their equipment as well as present a navigational hazard. Entanglement can damage fishing gear, lead to gear loss, or in extreme cases cause smaller fishing vessels to sink (Harrelson et al., 2022). To reduce both intentional and unintentional damage to UCH and increase the safety of captains operating in the area, SBNMS undertook a departure from the previous policy of non-disclosure in the Shipwreck Avoidance Pilot Program.

6 Multi-use and Multi-vocal Challenges of Preserving Underwater Cultural Heritage...   75

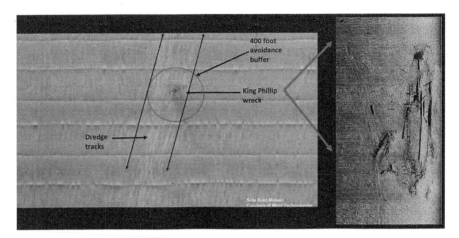

Fig. 6.5 Side-scan image of trawl marks through the possible *King Philip* shipwreck site (left); and through the ship's starboard bow section (right)

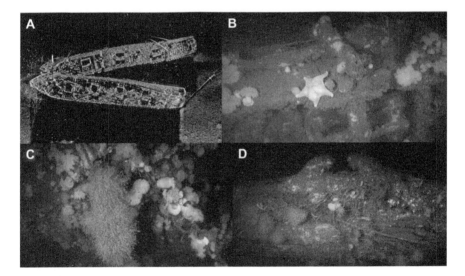

Fig. 6.6 Biological communities and impacts of fishing gear entanglement on the coal schooners *Frank A. Palmer* and *Louise B. Crary*. (**a**) sonar image of the two ships interlocked, courtesy of SBNMS; (**b**) a horse star (*Hippasteria phrygiana*) with fluffy anemones (*Metridium senile*) and sponges (*Mycale lingua*) on the shipwreck's hull; (**c**) dense clusters of hydroids (*Ectopleura crocea*) and fluffy anemones on monofilament strung between the hulls; (**d**) a net entangled on a cleat, with hydroids and sponges

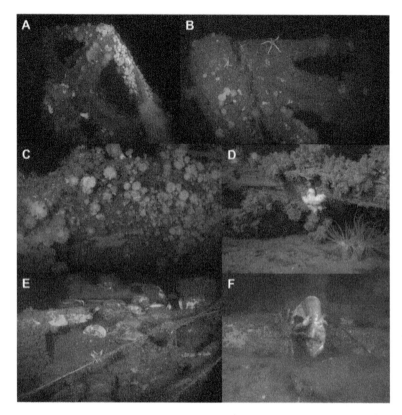

Fig. 6.7 Biological communities and impacts of fishing gear entanglement on *Portland*. (**a**) a dense population of fluffy anemones (Metridium senile) with entangled fishing gear (bottom left) on the walking beam; (**b**) a community of anemones and sea stars (Henricia sp.) on the walking beam; (**c**) fluffy anemones and sponges (Halichondria sp., Mycale lingua) on the aft hull; (**d**) tunicates (Molgula sp.) and tubed anemones (*Cerianthus borealis*) on a low-lying part of the shipwreck; (**e**) Acadian redfish (*Sebastes fasciatus*) rest on a pile of dinnerware near entangled fishing gear on the starboard side; (**f**) a stone crab (*Lithodes maja*) feasts on a cusk (*Bromse bromse*) that died following entanglement in ghost fishing gear. (All images courtesy of Marine Imaging Technologies)

6.5 Shipwreck Avoidance Pilot Program (SAPP)

In 2018–2021, SBNMS collaborated with NOAA fisheries to implement SAPP. The program involved the disclosure of several shipwreck locations at the beginning of the fishing season along with guidelines for voluntary avoidance with a 110-m buffer zone (Fig. 6.8). In 2018, SBNMS disclosed the locations of 3 modern and 4 historic shipwrecks for the first time in the history of the sanctuary. In 2020, the program expanded to include 4 more shipwrecks and an increased avoidance buffer of 122 m.

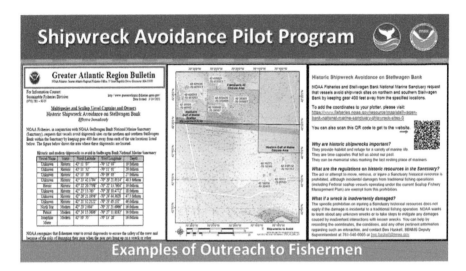

Fig. 6.8 Outreach bulletin requesting voluntarily avoidance of listed wreck sites (left); placards that could be placed in wheelhouses as avoidance reminders (middle and right); a QR Code (middle of right-hand panel) takes fishers to a NOAA Fisheries Web site where they can see the site locations and download coordinates to upload into their chart plotters

Data collection for the SAPP included mapping vessel tracks, pre- and post-season side- scan surveys, and interviews with fishing captains. The Automatic Identification System (AIS) and Vessel Monitoring System (VMS) were used to track vessel locations; however, these methods did not provide adequate resolution and coverage to be useful due to technology constraints involving time stamps (VMS) and vessel size requirements (AIS).

Interviews were conducted with 78 scallop, groundfish, and lobster vessel captains to assess the effectiveness of SAPP. Many vessel captains were supportive of the disclosure of shipwreck sites, although most local fishers were already aware of the locations. Mobile gear captains reported that they usually stayed more than 110 m away to reduce gear loss risk. However, some captains, particularly gillnet fishers, admitted to targeting wrecks. Side-scan sonar surveys proved to be the most effective tool for evaluating compliance of the voluntary avoidance buffers. The side-scan showed that compliance was incomplete due to evidence of continual activities (Harrelson et al., 2022).

There was interest among vessel captains to learn more about the shipwreck sites, particularly the human stories associated with them. Some captains felt it was unclear why particular wrecks were being preserved and stated that more historical information could impact their willingness to comply with voluntary avoidance guidelines. Most captains (53%) viewed the sites as having broad historic value or memorial significance, while some (21%) also named the creation of habitat to support fish populations (Harrelson et al., 2022). Some captains had never considered shipwrecks as valuable and thought of them primarily as hazards. Nevertheless, they viewed education about UCH to be an important next step to create a transparent process and develop voluntary guidelines for shipwreck avoidance.

The SAPP illustrated the challenges of protecting UCH in a multi-use environment. Compliance with voluntary avoidance buffers was mixed, and outreach to the fishing industry was challenging. The current sanctuary regulations provide insufficient deterrence for captains to avoid shipwrecks. Some captains also remain unconvinced that shipwreck sites on Stellwagen Bank need protection, although there was broad support for the continued release of wreck locations to aid in vessel safety. Expanded background information that helps illustrate the importance of shipwrecks, both to the sanctuary mission and to captains' values, could be a useful addition to future notices. Because formal sanctuary efforts to assess and protect important maritime heritage resources are still very much in development, there is a clear need for continued research to help guide best practices (Harrelson et al., 2022).

6.6 Integrating Maritime Heritage Ecology

The fishing captains' attitudes towards the perceived benefits of preserving shipwrecks and their desire for more information are emblematic of the multi-use and multivocality challenge SBNMS faces with its UCH and current policies. More than twice as many captains recognised the historical value of UCH than the ecological and biodiversity value. The third most common response was 'no benefit' (13%), indicating that fishers perceive *only* two benefits of UCH. Furthermore, only 3 of 78 respondents (<5%) perceived any benefit to researching shipwrecks (Harrelson et al., 2022). However, UCH offers many opportunities to understand ocean processes, and its archaeological and ecological roles are intertwined.

Therefore, SBNMS has adopted Maritime Heritage Ecology (MHE) to help shape their management efforts. MHE is an 'interdisciplinary research framework that aims to understand the interactive biological, natural, and anthropogenic factors that drive site formation processes and answer critical management questions for UCH' (Meyer-Kaiser & Mires, 2022). This framework can help integrate the disparate UCH contexts discussed here and hopefully help balance the dual mandate in SBNMS.

For example, the interdisciplinary framework of MHE could help integrate sociological and ecological factors to understand fishing patterns and engage fishers in active protection of key habitats—not just compliance with policies and regulations. Educating them about the importance of UCH habitats for vulnerable species such as cusk (*Bromse bromse*) and the endangered wolf fish (*Anarhichas lupus*) could help motivate shipwreck avoidance. Understanding risk-taking behaviours among fishers could also help managers craft incentive programs and ensure policy compliance. Furthermore, sharing of information about UCH, which fishers themselves suggested, will help preserve fishing heritage. The *North Star*, damaged by scallop fishing in 2017, was a steel clam dredge vessel and one of 45 shipwrecks representing the fishing industry in SBNMS. Engaging fishers in the process of heritage and embracing its multivocality will create a more personal connection to UCH and willingness to preserve these non-renewable resources.

MHE also provides an opportunity to engage other disciplines and technologies for innovative cooperation. For instance, moorings placed on UCH could serve as visual and digital markers of prescribed buffer zones and provide data on real-time conditions at sea. Similar programs already exist within the NMS in Michigan, Wisconsin, and Florida. Moorings become a part of the ecosystem, attract further biodiversity, and provide an anchor point for diving and swimming (Harrelson et al., 2022). Additionally, Massachusetts has a highly active shark monitoring program along the Cape Cod National Seashore, but there is little tracking information in the sanctuary and northwards. UCH could be used to provide data stations for these apex predators to help protect wildlife.

Anthropogenic structures are integral to marine ecosystems. MHE stresses that to protect biodiversity and preserve UCH, a clear understanding of all natural, biological, and anthropogenic processes impacting UCH is necessary. By engaging the fishing community in all components of MHE, SBNMS could build personal connections to the multivocal heritage and ecology of UCH. Interdisciplinary collaboration is needed for healthy fisheries and heritage preservation alike.

6.7 Conclusion

SBNMS is a case study in the multiutility and multivocality of heritage. The use and non-use values of UCH are defined differently by researchers, managers, and stakeholders, creating dissonance in preservation and management (Graham et al., 2000). Education and research in MHE can help build a common foundation and engage the fishing community. Management policies affect the livelihoods of fishers, so it is essential to consider their perspectives. This chapter has focused on fishing because it is the major management challenge in SBNMS, but a comprehensive management plan will engage other user groups, including SCUBA divers and indigenous tribes.

It is clear based on recent research that entanglement of trawl nets, monofilament, and other fishing gear is a major threat to shipwrecks in SBNMS. Protecting archaeological sites and biodiversity alike will require a radius around highly vulnerable UCH free from fishing activity. However, fishing is an economically important industry that provides identity for the Bay State. SBNMS is forced to balance these two competing priorities.

While a moratorium on fishing may be appropriate for NOAA Fisheries authorities to consider, the sanctuary lacks the authority to implement one without going through the entire sanctuary designation process again. It is also important for the sanctuary to avoid becoming a 'paper park' that fails in its conservation goals (Relano & Pauly, 2023). Shipwreck protection practices that have been developed in Thunder Bay and Florida Keys National Marine Sanctuaries provide a precedent for SBNMS. The SAPP made great strides in establishing communication with the fishing community and testing the feasibility of voluntary compliance. Expanding this program and exploring technological and infrastructure solutions for shipwreck avoidance will be key components of SBNMS management moving forward (SBNMS, 2023a).

Finally, expanding MHE research in SBNMS is critical to protecting resources. The newly implemented Sanctuary Mapping Initiative (SMI) enlists the help of fishers to conduct side-scan sonar surveys to locate and document shipwrecks and characterise seafloor habitats. Observational and experimental research on biological communities could provide new information on biodiversity hotspots and the role of UCH in supporting fisheries. Interviews and vessel monitoring will show how attitudes and behaviours of fishers affect site formation of UCH. The integrative, interdisciplinary approach of Maritime Heritage Ecology provides a foundation for SBNMS to balance the sanctuary's dual mandate and become an effective multi-use sanctuary for generations to come.

Acknowledgments The authors would like to thank all staff at Stellwagen Bank National Marine Sanctuary, particularly Clea Harrelson and Hannah McDonald; Marine Imaging Technologies; and the Ocean Vision 2030 Fund and its generous donors through the Woods Hole Oceanographic Institution.

References

Brennan, M. (2016). Quantifying impacts of trawling to shipwrecks. In M. E. Keith (Ed.), *Site formation processes of submerged shipwrecks* (pp. 157–179). University Press of Florida.

Brundage, W. F. (Ed.). (2000). *Where these memories grow: History, memory, and southern identity*. University of North Carolina Press.

Flatman, J. (2003). Cultural biographies, cognitive landscapes and dirty old bits of boat: 'Theory' in maritime archaeology. *International Journal of Nautical Archaeology, 32*(2), 143–157. https://doi.org/10.1111/j.1095-9270.2003.tb01441.x

Gould, R. (Ed.). (1983). *Shipwreck anthropology*. University of New Mexico Press.

Graham, B. J., Ashworth, G. J., & Tunbridge, J. E. (2000). *A geography of heritage: Power, culture, and economy*. Arnold/Oxford University Press.

Gravina, M. F., Casoli, E., Donnarumma, L., Giampaoletti, J., Antonelli, F., Sacco Perasso, C., & Ricci, S. (2021). First report on the benthic invertebrate community associated with a bronze naval ram from the first Punic war: A proxy of marine biodiversity. *Frontiers in Marine Science, 8*, 772499. https://doi.org/10.3389/fmars.2021.772499

Hamdan, L. J., Hampel, J. J., Moseley, R. D., Mugge, R. L., Ray, A., Salerno, J. L., & Damour, M. (2021). Deep-sea shipwrecks represent Island-like ecosystems for marine microbiomes. *The ISME Journal, 15*(10), 2883–2891. https://doi.org/10.1038/s41396-021-00978-y

Harrelson, C., MacDonald, H., Dalton, T., & Haskell, B. (2022). *Assessing efforts to mitigate fishing gear impacts on shipwreck sites in Stellwagen Bank National Marine Sanctuary* (ONMS-22-01; National Marine Sanctuaries Conservation Science Series). Office of National Marine Sciences, National Oceanic and Atmospheric Administration.

Jameson, J. H., & Scott-Ireton, D. A. (2007). Introduction: Imparting values/making connections. In J. H. Jameson & D. A. Scott-Ireton (Eds.), *Out of the blue: Public interpretation of maritime cultural resources* (pp. 1–6). Springer.

Kristiansen, K. (2000). Perspectives on archaeological heritage: History and future. In H. Cleere (Ed.), *Archaeological heritage management in the modern world* (pp. 23–29). Routledge.

Lawrence, M., & Marx, D. (2011). *Stellwagen Bank National Marine Sanctuary: 2010 maritime heritage ROV characterization project report*. Stellwagen Bank National Marine Sanctuary.

Lawrence, M., Marx, D., & Galluzzo, J. (2015). *Shipwrecks of Stellwagen Bank: Disaster in New England's national marine sanctuary*. The History Press.

Lipe, W. (2002). Public benefits of archaeological research. In B. Little (Ed.), *Public benefits of archaeology* (pp. 20–20). University of Florida Press.

Lowenthal, D. (1996). *Possessed by the past: The heritage crusade and the spoils of history*. Free Press.

McManamon, F. (2002). Heritage, history, and archaeological educators. In B. Little (Ed.), *Public benefits of archaeology* (pp. 31–45). University of Florida Press.

Meyer-Kaiser, K. S., & Mires, C. H. (2022). Underwater cultural heritage is integral to marine ecosystems. *Trends in Ecology & Evolution, 37*(10), 815–818. https://doi.org/10.1016/j.tree.2022.06.014

Meyer-Kaiser, K. S., Mires, C. H., & Haskell, B. (2022a). Invertebrate communities on shipwrecks in Stellwagen Bank National Marine Sanctuary. *Marine Ecology Progress Series, 685*, 19–29. https://doi.org/10.3354/meps13987

Meyer-Kaiser, K. S., Mires, C. H., Kovacs, M., Kovacs, E., & Haskell, B. (2022b). Structural factors driving benthic invertebrate community structure on historical shipwrecks in a large North Atlantic marine sanctuary. *Marine Pollution Bulletin, 178*, 113622. https://doi.org/10.1016/j.marpolbul.2022.113622

Mires, C. H., & Meyer-Kaiser, K. S. (2023). A case study in maritime heritage ecology: Understanding how structural changes to the 1898 shipwreck Portland affect biological diversity and colonization. *Journal of Maritime Archaeology, 18*(2), 197–218. https://doi.org/10.1007/s11457-023-09359-2

Mires, C., Kovacs, E., & Meyer-Kaiser, K. (2020). Return to Portland 2019: Stellwagen Bank National Marine Sanctuary and Telepresence. In V. Mastone & C. Mires (Eds.), *ACUA underwater archaeological proceedings 2020* (pp. 101–109). The PAST Foundation.

Murphy, L. (1983). Shipwrecks as a data base for human behavioural studies. In R. Gould (Ed.), *Shipwreck anthropology*. University of New Mexico Press.

Office of National Marine Sanctuaries (ONMS). (2020). *Condition report: Findings of status and trends for 2007–2018*. U.S. Department of Commerce, National Oceanic and Atmospheric Administration.

Osborne, B. (1998). Constructing landscapes of power: The George Etienne Cartier monument, Montreal. *Journal of Historical Geography, 24*, 431–458.

Paxton, A. B., Peterson, C. H., Taylor, J. C., Adler, A. M., Pickering, E. A., & Silliman, B. R. (2019). Artificial reefs facilitate tropical fish at their range edge. *Communications Biology, 2*(1), 168. https://doi.org/10.1038/s42003-019-0398-2

Perkol-Finkel, S., & Benayahu, Y. (2007). Differential recruitment of benthic communities on neighboring artificial and natural reefs. *Journal of Experimental Marine Biology and Ecology, 340*(1), 25–39. https://doi.org/10.1016/j.jembe.2006.08.008

Perkol-Finkel, S., Shashar, N., & Benayahu, Y. (2006). Can artificial reefs mimic natural reef communities? The roles of structural features and age. *Marine Environmental Research, 61*(2), 121–135. https://doi.org/10.1016/j.marenvres.2005.08.001

Pile, S., & Keith, M. (Eds.). (1997). *Geographies of resistance*. Routledge.

Price, K. A., Garrison, C. E., Richards, N., & Field, E. K. (2020). A shallow water ferrous-hulled shipwreck reveals a distinct microbial community. *Frontiers in Microbiology, 11*, 1897. https://doi.org/10.3389/fmicb.2020.01897

Relano, V., & Pauly, D. (2023). The 'Paper Park Index': Evaluating marine protected area effectiveness through a global study of stakeholder persceptions. *Marine Policy, 151*, 105571.

Rosenzweig, R., & Thelen, D. (1998). *The presence of the past: Popular uses of history in American life*. Columbia University Press.

Ross, S. W., Rhode, M., Viada, S. T., & Mather, R. (2016). Fish species associated with shipwreck and natural hard-bottom habitats from the middle to outer continental shelf of the Middle Atlantic Bight near Norfolk Canyon. *Fishery Bulletin, 114*(1), 45–57. https://doi.org/10.7755/FB.114.1.4

Seaton, A. (2001). Sources of slavery-destination of slavery: The silences and discourses of slavery heritage in the UK and US. In G. Dann & A. V. Seaton (Eds.), *Slavery, contested heritage, and thanatourism* (pp. 107–129). Haworth Hospitality Press.

Siciliano, A., Jimenez, C., & Petrou, A. (2016). Recreational diving and its effects on the macroalgal communities of the unintentional artificial reef Zenobia shipwreck (Cyprus). *Journal of Oceanography and Marine Research, 04*(02). https://doi.org/10.4172/2572-3103.1000151

Stellwagen Bank National Marine Sanctuary (SBNMS). (2021). *Stellwagen Bank National Marine Sanctuary draft management plan and environmental assessment*. Office of National Marine Sanctuaries, National Oceanic and Atmospheric Administration.

Stellwagen Bank National Marine Sanctuary (SBNMS). (2023a). In Office of National Marine Sanctuaries, National Oceanic and Atmospheric Administration (Ed.), *Stellwagen Bank National Marine Sanctuary draft management plan and environmental assessment*.

Stellwagen Bank National Marine Sanctuary (SBNMS). (2023b). *Stellwagen Bank National Marine Sanctuary shipwrecks*. https://stellwagen.noaa.gov/maritime/shipwrecks.html

Trigger, B. G. (2009). *A history of archaeological thought* (2nd ed., repr). Cambridge University Press.

United Nations Educational, Scientific and Cultural Organization (UNESCO). (2001). *Convention on the protection of the underwater cultural heritage*. https://www.unesco.org/en/underwater-heritage. Accessed 4 April 2024.

Watson, P. J. (1983). Method and theory in shipwreck archaeology. In R. Gould (Ed.), *Shipwreck anthropology* (pp. 23–36). University of New Mexico Press.

Open Access This chapter is licensed under the terms of the Creative Commons Attribution 4.0 International License (http://creativecommons.org/licenses/by/4.0/), which permits use, sharing, adaptation, distribution and reproduction in any medium or format, as long as you give appropriate credit to the original author(s) and the source, provide a link to the Creative Commons license and indicate if changes were made.

The images or other third party material in this chapter are included in the chapter's Creative Commons license, unless indicated otherwise in a credit line to the material. If material is not included in the chapter's Creative Commons license and your intended use is not permitted by statutory regulation or exceeds the permitted use, you will need to obtain permission directly from the copyright holder.

Chapter 7
Underwater Cultural Heritage and Fishing Communities: Safeguarding Heritage and Safeguarding Fishers

Benjamin Ferrari, Antony Firth, David Gregory, and Louise Sanger

7.1 Introduction

This paper proposes that three elements are required in order to develop a fit-for purpose management regime to protect underwater cultural heritage (UCH) from negative impacts caused by some fishing gear: evidence-based characterisation of impacts; understanding of the pressures and constraints acting upon fishing communities; and commitment to engagement with fishers and industry associations. In addition, strategies for protection of UCH from fishing and other hazards cannot be developed in isolation from existing marine management regimes. This challenge of integration and alignment with broader stewardship of complex marine systems is set to grow more difficult as industrial activity grows, competition for resources intensifies, and governments and agencies seek to exert new levels of control over larger areas of ocean.

There are more than three million shipwrecks estimated worldwide according to UNESCO (Croome, 1999). The UNESCO convention on the protection of underwater cultural heritage, encourages *in situ* preservation (UNESCO, 2001) and states that wrecks older than 100 years count as a cultural heritage site. This definition now encompasses many metal shipwrecks, including World War I casualties and

---

B. Ferrari (✉) · L. Sanger
Lloyd's Register Foundation Heritage Education Centre, London, UK
e-mail: Benjamin.Ferrari@lr.org

A. Firth
Historic England, London, UK

D. Gregory
Department of Research, Collections and Conservation, Section for Environmental Archaeology and Materials Science, National Museum of Denmark, Copenhagen, Denmark

© The Author(s) 2024
C. Jarvis (ed.), *Threats to Our Ocean Heritage: Bottom Trawling*,
SpringerBriefs in Underwater Archaeology, https://doi.org/10.1007/978-3-031-57953-0_7

soon will encompass casualties from World War II. Significant conflicts of interest are now evident—even in the context of conservation-oriented assessments (Firth, 2018). Shipwrecks are habitats for marine life, enhancing biodiversity and biomass (Balazy et al., 2019). They can also pose environmental pollution threats, especially if they contain hazardous substances such as oil, chemicals, or munitions (Ndungu et al., 2017; Szafrańska et al., 2021). Indeed, this issue of potentially hazardous and polluting legacy wrecks is receiving focused attention from influential advisory bodies. The International Union for the Conservation of Nature (IUCN) recently passed resolutions urging its members to take action with a view to preserving biodiversity (IUCN, 2020). The challenge of addressing the urgent need for remediation while also acknowledging heritage value (including potential war grave status) will need to be addressed comprehensively. Marine spatial planning issues often arise as wrecks can interfere with the ever-increasing offshore sub-sea engineering efforts, such as the installation of cables, pipelines, or wind farms with a need to protect heritage value—a source of potential added costs to developers and operators (Papageorgiou, 2018). Shipwrecks are also treated as geohazards and obstacles to marine engineering, posing risks to navigation or safety.

The impact of fishing activity on UCH must be addressed but such impacts can also be a major concern to owners and operators of ocean infrastructure who have proposed various forms of fishing exclusion zones and technical remedies. Therefore, while UCH and its management, as described above, can give rise to conflicts of interest, there are also shared concerns between heritage managers and other stakeholders in the context of a complex marine management and stewardship environment.

The first section of this chapter highlights the need for evidence-based characterisation of the impact of fishing and integration of this analysis with wider consideration of site formation dynamics. A major European Research Council research project, ENDURE, is introduced. Based at the National Museum of Denmark, ENDURE is a multi-year programme that aims to disentangle natural and anthropogenic decay processes and determine their cumulative effects on UCH. The aim is to validate the efficacy of *in situ* preservation.[1] The subsequent sections develop the proposition that the successful application of such foundational work on new management regimes will be compromised in the absence of full appreciation of the pressures and constraints operating on the fishing industry in general and coastal fisheries in particular. Effective management of UCH requires concerted effort committed to close engagement with fishers. Not all fishing activity poses an equal threat to UCH, and it is critical that an understanding of material differences in impact influences the management solutions proposed.

The second section introduces the work of Lloyd's Register Foundation (LRF), shaping a future for the fishing industry that is safe for all fishers working in a sustainable ocean economy. Today's fisheries are part of a highly stressed marine system, and the human cost of fishing is appallingly high. LRF's work will help inform

---

[1] Project ENDURE https://www.endureerc.com

proposals for protection of UCH that avoid exacerbating occupational safety challenges. The third section presents work undertaken by Historic England (HE) to engage with fishers in order to mitigate damage to UCH and develop fit-for-purpose management regimes. It emphasises the need for broad engagement with agencies as well as industry bodies and communities.

7.2 Out of Sight but Not Out of Mind: Sustainable Preservation of Underwater Archaeological Sites

The scale of the heritage management challenge is considerable with a wide range of site types in complex, dynamic environments. Consequently, a desk-based approach for identification and assessment of the most significant site formation processes is urgently needed especially as rapidly progressing marine remote sensing techniques allows effective non-intrusive detection, characterisation, and monitoring of underwater sites (Plets et al., 2011; Westley et al., 2019). The remote characterisation and monitoring of change at individual sites (Quinn & Boland, 2010; Brennan et al., 2016) includes environmental factors to understand the physical formation processes that are most influential in determining levels of site preservation (Smyth & Quinn, 2014; Fernández-Montblanc et al., 2018; Quinn & Smyth, 2018; Geraga et al., 2020; Majcher et al., 2021). Gregory and Manders (2015) formed a baseline process-based approach to control, assess and achieve *in situ* preservation.

One of its central points is the assessment of the most significant chemical, biological and physical threats to the site. Understanding the formation of underwater sites also enables more accurate archaeological interpretation. Assessing environmental factors can not only lead to better understanding of events that caused the sinking of the vessel but also help to predict pathways of future deterioration. This enhances understanding of potential pollution and fishing hazards and addresses UCH management needs.

In ongoing work (Gregory et al., 2024) baseline post-depositional site formation processes for more than 500 shipwreck sites, located in the North and Baltic Seas, are being examined. Multi-beam bathymetry data supplied by a marine survey company (JD-Contractor A/S) are integrated with multiple environmental and anthropogenic influence data layers sourced from open-data portals. A revised form of Muckelroy's (1977) approach was used to achieve an initial manual classification of the sites and subsequently, statistical correlation analysis was used to determine the most significant variables. Fishing activity was assessed through inclusion of variables representing fishing efforts and only the fishing methods entailing direct contact of the fishing gear with the seabed were deemed to be relevant in this study. EMODnet (2022) provides a variety of quantified fishing activities in the Baltic and

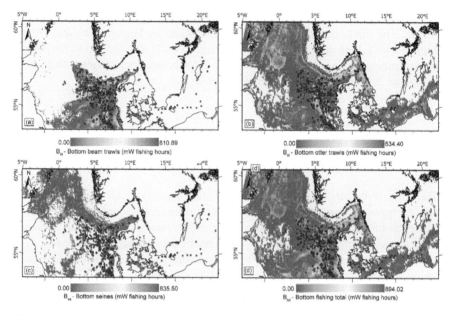

Fig. 7.1 Fishing intensities. (**a**) Bottom beam trawling, (**b**) bottom otter trawling, (**c**) bottom seines, (**d**) total bottom fishing intensity. Grey areas represent lack of data. (Figure from Gregory et al. (2024) prepared by Jan Majcher)

North Seas and bottom seiner, beam and otter trawler fishing data were downloaded from the EMODnet portal, representing fishing intensities averaged across three years (Fig. 7.1). However, some caution is required as there is on-going research into the footprint and actual impact of fishing on seabed habitats that may allow further refinement of this methodology (McConnaughey et al., 2020).

Although the study presents a simplified, preliminary approach, it serves as a big-data-driven attempt to assess the predominant basin-scale post-depositional site formation factors acting at UCH sites in order to enable strategic, evidence-based management decisions to be made. Water depth, bottom fishing intensity and oceanographic variables like salinity have significant influence in addition to the expected correlation between the age of a wreck site (sinking date) and its current preservation state. These aspects are being further developed in the ENDURE project addressing the following key knowledge gaps: (1) the efficacy and long term sustainability of '*in situ* preservation'; (2) what to preserve and why; (3) a comprehensive understanding of decay processes (including their rates) and how and in what circumstances these can be mitigated for; (4) the invisible and inaccessible nature of these sites and how such a large resource can be effectively monitored; (5) the increasingly dynamic and changing nature of the marine environment due to both natural and anthropogenic drivers; and (6) the lack of integration of UCH management with other disciplines and end users of the marine environment.

7.3 The World's Most Dangerous Industry

The damage caused to UCH by trawling activity cannot be ignored by responsible Government agencies and Lloyd's Register Foundation[2] (LRF) is committed to supporting the safe and effective management of UCH. However, LRF believes that attempts to develop better management regimes for UCH located on or near fishing grounds must be informed by a high-level of awareness of challenges faced by the fishing community—especially in the global south. Accordingly, LRF Heritage Education Centre, has launched the *Learning from the Past* programme (LRF n.d.-a). The aim is to use historical and archaeological evidence to generate insights that contribute to contemporary fisher safety while also helping archaeologists and historians to communicate with fishers and policy makers in order to enhance efforts to limit harm to both the natural and historic marine environments.

Illegal fishing, climate change, weak governance, poverty and reduced fish stocks all contribute to make fishing one of the world's most dangerous professions. Recent work by the FISH Safety Foundation indicates that more than 100,000 fishing-related fatalities occur annually with the majority in the developing world (Willis et al., 2023; FISH Safety Foundation, 2023). Negative health effects on industry participants in general and fatalities greatly exceed those in the wider shipping sector (S&P Global, 2020). In Bangladesh, more than 1350 fishers die at sea every year, often in boats lacking the most basic safety equipment and these losses can result in profound hardship for families and communities (FISH Safety Foundation, 2023).

Lloyd's Register Foundation identified fishing as a safety challenge in its highly influential *Insight Report on Safety in the Fishing Industry* (Attwood, 2018). Since then, the Foundation has further investigated to better understand the nature of the safety challenge in the fishing industry. It has drawn on expert knowledge and opinion to understand what activity is already underway to improve fishing safety and what additional action is needed. Causes of fatalities differ in the developed and developing world. However, they are generally related to the economic status, value placed on life, and social outlook of the fishers in the respective countries. The report recommended that an international programme should be established, focused on local community partnerships, to improve safety in ways that are appropriate for local communities. LRF believes that this community approach is a critical element of managing fishery impacts on UCH. LRF has subsequently launched several initiatives, such as FishSafe2025, to help assure fisher safety through training and awareness (LRF, 2020). The Foundation has also supported development of new safety technology—for example SeaWise®, a tablet-sized device that monitors the stability of smaller fishing vessels (LRF, 2021). These outputs contain many insights that may be valuable in designing new UCH management regimes—especially in inshore waters.

Marine Protected Areas (MPAs) and latterly, Highly Protected Marine Areas (HPMAs), will be a key tool in both marine nature and heritage conservation.

---

[2] https://www.lrfoundation.org.uk/en/

However, the resulting displacement of fishers from familiar grounds, whether through creation of MPAs or other processes, can carry a severe cost that must be recognised and addressed in planning stages. The LRF funded film *Two Kinds of Water* tells the story of fishers in Africa's most vulnerable fishing communities (LRF, 2022). Displaced from their traditional fishing grounds into stormier waters, fishers are forced to venture further and stay at sea longer in poorly adapted vessels. The need to operate further from land also results in loss of access to mobile telecommunication networks—previously the prime source of weather and other safety information as well as contact with the shore. Larger, mechanised fishing vessels, also moving from regular fishing grounds due to reduction and migration of fishstocks due to overfishing, now compete with artisanal fishers for fish and sea-space. They pose a direct safety threat to the more fragile, smaller craft and mortality rates are rising as catches fall.

This is not a problem confined to the Global South. Fishing communities around the UK face disruption and displacement. Fishers have been moved from traditional shore facilities due to redevelopment causing operational and occupational safety issues. The UK Government plans for new HPMAs may also deny fishers access to traditional inshore grounds (DEFRA, 2019). UK sea fisheries associations have warned of a looming displacement crisis with increased risk and lasting harm to coastal communities (National Federation of Fishermen's Organisations, 2022).

There have been extensive consultations on proposed HPMAs off the English coast and eventually two of the proposed designations (Lindisfarne and Inner Silver Pit South) were refused. The needs of fishing communities were prominent in the published rationale for rejection (DEFRA, 2019). For Lindisfarne, reasons included residents' concerns about losing their heritage, community and cultural identity through losing fishing and also their health and safety concerns, including mental health. The Inner Silver Pit South designation was refused on the basis that the relatively high costs to fishers would not be offset by the potential benefits from its designation as a HPMA.

A similar pattern has emerged in Scotland (in which, like in England, relevant powers are devolved to the Scottish Parliament). The initial announcements by Scottish Ministers included commitments to designate at least 10% of Scotland's seas as Highly Protected Marine Areas by 2026 (Scottish Government, 2022). However, there was an immediate backlash that united coastal communities, grassroots campaigners and sectors of the fishing industry—even spawning a popular protest song (SKIPINNISH, 2023). Former Finance Secretary Kate Forbes warned people, not wildlife, could become the 'endangered species' in Highland and Island areas due to jobs being lost as a result of the proposals. The strength of the opposition has resulted in withdrawal of the initial timetable for creation of HPMAs and the First Minister of Scotland, Humza Yousaf, stated that no HPMAs would be imposed on island and coastal communities without their consent (HPMAs, 2023).

7.4 Protecting England's Maritime Heritage and Fishing Community Engagement

Fishing in English waters, using both mobile and static gear, involves a variety of complex interactions with underwater cultural heritage. Fishing has given rise to important heritage both underwater and along our coastlines, but fishing can also cause damage to UCH. Fishing is often a source of information about UCH—in the form of new discoveries—even where those discoveries are a consequence of fishing impacts.

Historic England (HE) is the government's advisor on the historic environment in England, including in the UK Marine Area (UKMA) off England,[3] encompassing the territorial sea to 12 nautical miles (nm), and the UK Continental Shelf and Exclusive Economic Zone to median lines with neighbouring countries or to the boundary of national jurisdiction (200 nm) in the southwest. HE's capacity to designate sites under heritage legislation is limited to the territorial sea but HE advises government on a range of other powers relating to resource use, the environment and fishing that apply to the full extent of national jurisdiction. For fisheries management, another key boundary lies at six nautical miles, which is the outer limit of areas administered by Inshore Fisheries and Conservation Authorities (IFCAs). Beyond six nautical miles, fisheries management was subject to the Common Fisheries Policy (CFP) of the European Union (EU). Since the UK left the EU, fisheries off England beyond six nautical miles are now administered by the Marine Management Organisation (MMO), one of the family of public bodies sponsored by the Department for Environment, Food and Rural Affairs (DEFRA).

HE, including through its precursor English Heritage, has been engaging with fisheries for many years. It commissioned a wide-ranging examination of fishing and the historic environment that encompassed the contribution of fishing to the historic environment as well as direct and indirect interactions, both positive and negative (Firth et al., 2013). The report spurred further investigations and continues to provide evidence to support advice to government.

In 2012–2013, HE funded a pilot project to encourage fishers to report their discoveries of UCH through a Fishing Industry Protocol for Archaeological Discoveries (FIPAD) (Davidson, 2013). FIPAD was further developed with the support of the National Lottery Heritage Fund and Sussex IFCA in 2016–2018.[4]

At an earlier juncture—in the 1990s—'fishermen's fasteners' were among the core data recorded in a national inventory now known as the National Marine Heritage Record. Fasteners are places where fishers have noted snags to their gear which, in several cases, proved to be historic wrecks. Whilst fasteners indicate potential UCH, their actual character is unknown. This, and several other ambiguities, prompted a project led by Sussex IFCA to better understand fasteners and their

---

[3] Cultural heritage and fisheries in the UK are both devolved matters: distinct arrangements apply in England, Scotland, Wales, and Northern Ireland.
[4] https://fipad.org/

heritage implications, including testing whether the survey methods used by IFCAs for fisheries and habitat mapping could also be used to examine fasteners. The project, which incorporated dialogue and joint fieldwork with fishers, confirmed the continued importance of fishermen's fasteners as a source of information (Firth & Dapling, 2020). The project also flagged the safety issues presented by cultural heritage to fishing insofar as some fishing vessels, and fishers' lives, have been lost from snagging obstructions and other historic material such as ordnance (Marine Accident Investigation Branch, 2015; Marine Accident Investigation Branch, 2022). The potential for IFCAs to use habitat survey methods to examine UCH was further explored in a project led by Isles of Scilly IFCA with the University of Plymouth (Firth et al., 2020).

Notwithstanding these positive instances of collaboration, fishing is still a cause of impacts to UCH, including to highly significant historic wrecks designated under heritage legislation. Several instances of documented damage are currently being investigated as heritage crimes and in other instances fieldwork has been commissioned to remove Abandoned, Lost and Discarded Fishing Gear (ALDFG) that is obscuring designated sites and presenting a hazard to licensed divers. Whilst damage is a criminal offence, fishing as an activity is not restricted on these sites. Consequently, and in addition to pursuing criminal damage, HE efforts are twofold: working with the fishing sector to discourage the use of potentially damaging methods in the vicinity of important sites; whilst also exploring the use of fisheries management measures to formally restrict fishing.

New fisheries legislation, the Fisheries Act 2020, was introduced to accompany the UK leaving the EU and encompasses 'features of archaeological or historic interest' within the scope of the marine and aquatic environment that can be conserved, enhanced, and restored through financial assistance, regulations, or conditions on sea fishing licences. It was confirmed by the Minister in Parliament that regulations for conservation purposes may be used to amend or introduce legislation to protect features of archaeological or historic interest individually or collectively (Fisheries Bill [*Lords*], 2020, column 144).

Exclusion of fishing activity from limited areas around specific heritage sites is a residual option if the risk of impact cannot be reduced through other means, but it is not the starting point. As noted earlier, fishing is a valued part of our heritage, contributing to the character and sense of place of coastal communities, including people who work in them or visit. Often, UCH is itself the heritage of fishing generally but also of the fishers themselves: numerous fishing vessels lie wrecked in English waters, not least from the First and Second World Wars when fishing vessels and fishing crews served in the hazardous role of minesweeping. HE has supported the contribution of fishing heritage to regeneration of coastal communities through Heritage Action Zones (HAZs) and High Street Action Zones (HAZs). Examples of this include the 'Kasbah' at the heart of Grimsby's historic fishing industry (Historic England, n.d.), and the Whapload Road area of North Lowestoft (Bristow, 2019). HE is carrying out a thematic assessment of heritage relating to steam fishing, encompassing heritage assets on land and at sea.

HE continues to explore how heritage contributes to the sustainability of fishing, reducing impacts to UCH but potentially contributing to the wider sustainability of our continued use and enjoyment of our seas. HE is supporting and co-supervising a PhD on using cultural heritage proactively to help manage UK marine fisheries in conjunction with Heriot Watt University and the University of Exeter as part of the Centre for Doctoral Training in in Sustainable Management of UK Marine Resources (CDT SuMMeR). The research will examine how the legacy of cultural heritage embedded in the practice of small-scale fishing—traditional ecological knowledge, maritime landscapes, historic landing places, traditional vessels, and all their associated skills and material culture—can be used to drive greater sustainability within fisheries management.

Historic England continues to engage with DEFRA and the MMO on major changes to fisheries management introduced by the Fisheries Act 2020. The new Joint Fisheries Statement (JFS)—which sets out the overall direction of fisheries management across the UK –acknowledges that the seafood sector is an important part of the economy of coastal communities and has a rich cultural heritage from which many of those communities draw a sense of place and identity (DEFRA et al., 2022). There is work to do, however, on detailing how this rich cultural heritage is to be conserved and enhanced through the JFS. Historic England is engaged on the practical implications of the Fisheries Management Plans (FMPs) that support the JFS through binding obligations and identified measures for dealing with individual fish stocks, types of fishing and geographic areas (DEFRA et al., 2023). The introduction and implementation of FMPs provides a key opportunity to flag the impacts of specific fisheries and how they can be reduced, but also the positive contribution that heritage can make to the long-term sustainability of fishing and coastal communities.

7.5 Conclusion

There is undoubtedly an urgent need for greatly improved protection of UCH from certain fishing gear impacts. However, this challenge, nested as it is within multiple, broader ocean stewardship issues, is not set to become any less complex—possibly quite the reverse.

The pace and scope of growth in the ocean-based economy has led to realisation that we are on the verge of a new maritime industrial revolution without an adequate governance framework or sufficient information to ensure that safety and sustainability can be assured. The notion of the Blue Economy is well established. Ocean industries underpin our critical infrastructures and supply chains—80% of our goods are transported by ship and it is estimated that $90 trillion will be invested over the next decade on marine infrastructure alone. Oceans contribute $1.5 trillion annually to the global economy, and this number is expected to double to $3 trillion by 2030 (LRF, 2021). The UN estimates that the ocean directly supports the livelihood of about 500 million people and many poorer countries are almost wholly dependent on the ocean economy. As the global population is estimated to grow to 8.5 billion by 2030 and 9.7 billion by 2050, there will be increased pressure on ocean resources such as increasing demand for food, energy, jobs, transportation, and coastal land.

There is a broad acceptance that the Blue Economy must be a sustainable one. A just transition to a low carbon, sustainable ocean economy necessitates investment, education, infrastructure, innovation, and decent, safe jobs. But the capacity to industrialise is growing ever more rapidly. Marine ecosystems face unprecedented cumulative pressures from human activities and climate change. This is happening in a largely unmapped geography and a complex and uncertain governance landscape—a phenomenon dubbed the 'Blue Acceleration' (Jouffray et al., 2020). Put simply, the ocean has never been busier: shipping has increased by 1600% since 1982, when the UN Law of the Sea Convention was signed and a phenomenal rate of change over the last 50 years has produced a race among diverse and often competing interests for ocean food, material, and space.

Marine systems are becoming increasingly complex and safe management of shared space at sea is becoming a central topic of concern. Current management effort is necessarily focused on actions within territorial jurisdictions. However, there is increasing interest in management regimes for resources in areas beyond national jurisdiction as competition for these resources escalates (United Nations, 2023). These areas comprise 95% of the ocean and UCH is widely present within them, in both tangible and intangible form. For example, alongside individual wreck sites of varying antiquity, The Middle Passage, a maritime heritage landscape in the Atlantic Basin of immense significance to the African diasporic cultural memory, is potentially vulnerable to industrial activity such as mineral extraction being considered for licensing by the International Seabed Authority (Turner et al., 2020).

There is a time-limited conjunction of threat and opportunity. Never has UCH been under greater threat from industrial activity and fishing activities. Equally, there has never been a better opportunity to define and promote the standard and protocols required to deal with these threats and safety challenges and to communicate them as part of broader, international efforts to create a safe and sustainable ocean economy. As described in the preface of this volume, The UN Decade of Ocean Science for Sustainable Development (2021–2030) was launched with the vision 'the science we need for the ocean we want'.[5]

The critical role of maritime heritage in achieving this outcome was recognised early, with the Cultural Heritage Framework Programme (CHFP), led by the Ocean Decade Heritage Network (ODHN), being among the first of the 'Actions' to receive formal UN Decade endorsement. The CHFP offers an efficient interface between heritage and the Ocean Decade, providing advice and assistance to other programmes, projects and activities in key areas.[6]

---

[5] https://oceandecade.org/

[6] https://oceandecade.org/actions/cultural-heritage-framework-programme/

It is essential to use this channel to foster ongoing dialogue with the broader ocean sustainability community based on an evidence-led approach to the actual nature and scale of impacts created by different types of fishing gear in different locations. Fishing is also an inherently hazardous occupation and any approach that fails to address impact on occupational safety will founder. The built heritage associated with coastal fisheries is increasingly recognised for its contribution to a 'sense of place valued by the broader community' (Khakzad & Griffith, 2016) as are the embedded intangible cultural heritage values widely noted in consultations and programmes supporting the regional socio-economic role of fisheries.[7] Heritage agencies have worked successfully with fishing communities to explore collaborative management approaches. Indeed, examples exist of fishing communities recruited and trained specifically to assist with monitoring and conservation of UCH. However, much more effort is needed to develop ways to balance the needs of fishers with preservation of nature and heritage. Lloyd's Register Foundation will continue to support such efforts (LRF, n.d.-b).

References

Attwood, D. (2018). *Insight report on safety in the fishing industry: A global safety challenge.* Lloyds Register Foundation. https://www.lrfoundation.org.uk/en/publications

Balazy, P., Copeland, U., & Sokołowski, A. (2019). Shipwrecks and underwater objects of the southern Baltic—Hard substrata islands in the brackish, soft bottom marine environment. *Estuarine, Coastal and Shelf Science, 225*, 106240. https://doi.org/10.1016/j.ecss.2019.05.022

Brennan, M. L., Davis, D., Ballard, R. D., Trembanis, A. C., Vaughn, J. I., Krumholz, J. S., Delgado, J. P., Roman, C. N., Smart, C., Bell, K. L. C., Duman, M., & DuVal, C. (2016). Quantification of bottom trawl fishing damage to ancient shipwreck sites. *Marine Geology, 371*, 82–88. https://doi.org/10.1016/j.margeo.2015.11.001

Bristow, M. (2019). *311–333 Whapload Road, Lowestoft, Suffolk: Historic area assessment* (57–2019; Research report series). Historic England. https://historicengland.org.uk/research/results/reports/7747/311%E2%80%93333WhaploadRoadLowestoftSuffolk%E2%80%93HistoricAreaAssessment

Croome, A. (1999). Sinking fast. *New Scientist, 161*(2169), 49–52.

Davidson, S. (2013). *Fishing industry protocol for archaeological discoveries: final report* (73271.04). Wessex Archaeology. https://historicengland.org.uk/research/results/reports/6952/FishingIndustryProtocolforArchaeologicalDiscoveries

Department for Environment, Food & Rural Affairs (DEFRA). (2019). *Highly Protected Marine Areas (HPMAs) review 2019* [Independent report]. https://www.gov.uk/government/publications/highly-protected-marine-areas-hpmas-review-2019

Department for Environment, Food & Rural Affairs, Department of Agriculture, Environment and Rural Affairs (Northern Ireland), The Scottish Government, & Welsh Government. (2022). *Joint Fisheries Statement (JFS)*. https://www.gov.uk/government/publications/joint-fisheries-statement-jfs

Department for Environment, Food & Rural Affairs, Welsh Government, Marine Management Organisation, & Seafish. (2023, July 17). *Fisheries management plans*. Collection. https://www.gov.uk/government/collections/fisheries-management-plans

---

[7] https://www.cabfishman.net

European Marine Observation and Data Network (EMODnet). (2022, March 30). *Fishing intensity data set extended to additional sea regions*. https://emodnet.ec.europa.eu/en/fishing-intensity-data-set-extended-additional-sea-regions

Fernández-Montblanc, T., Izquierdo, A., Quinn, R., & Bethencourt, M. (2018). Waves and wrecks: A computational fluid dynamic study in an underwater archaeological site. *Ocean Engineering, 163*, 232–250. https://doi.org/10.1016/j.oceaneng.2018.05.062

Firth, A. (2018). *Managing shipwrecks*. Honor Frost Foundation. https://honorfrostfoundation.org/wp-content/uploads/2019/07/BRIJ5800-Multiwreck-A4-Report-WEB-0419-UPDATE.pdf

Firth, A., & Dapling, T. (2020). Fishing for heritage: A project to better understand "fishermen's fasteners" off the coast of Sussex. *Historic Environment Research, 15*, 38–45.

Firth, A., McAleese, L., Smith, R., & Woodcock, T. (2013). *Fishing and the historic environment* (EH 6204). Marine Ecological Surveys Ltd.

Firth, A., Hooper, T., & Sheehan, E. (2020). Fish and ships: Integrating heritage with habitat surveys off the Isles of Scilly. *Historic Environment Research, 15*, 46–53.

FISH Safety Foundation. (2023). *The human cost of fishing*. https://www.the-human-cost-of-fishing.org/

Fisheries Bill [Lords]. (2020). https://hansard.parliament.uk/Commons/2020-09-15/debates/a7b42ae4-819e-45c7-a0d6-fc32b494aacf/FisheriesBill(Lords)(FifthSitting)#contribution-FCA4B1FA-E107-4602-B34C-E9AF6F80A30F

Geraga, M., Christodoulou, D., Eleftherakis, D., Papatheodorou, G., Fakiris, E., Dimas, X., Georgiou, N., Kordella, S., Prevenios, M., Iatrou, M., Zoura, D., Kekebanou, S., Sotiropoulos, M., & Ferentinos, G. (2020). Atlas of shipwrecks in Inner Ionian Sea (Greece): A remote sensing approach. *Heritage, 3*(4), 1210–1236. https://doi.org/10.3390/heritage3040067

Gregory, D., & Manders, M. (2015). *Best practices for locating, surveying, assessing, monitoring and preserving underwater archaeological sites*. SASMAP guideline manual 2. http://sasmap.eu/fileadmin/user_upload/temasites/sas_map/pdf/SASMAP_guideline_02_LR.pdf

Gregory, D., Dam, M., Majcher, J., Matthiesen, H., Andersen, G. N., & Quinn, R. (2024). Using open-data portals, remote sensing and computational modelling to investigate historic wreck sites and their environments: 45 Years on from Muckelroy. *International Journal of Nautical Archaeology*, 1–18. https://doi.org/10.1080/10572414.2024.2320774

Historic England. (n.d.). *Greater Grimsby Heritage Action Zone*. Services & Skills: Heritage Action Zones. https://historicengland.org.uk/services-skills/heritage-action-zones/greater-grimsby/

HPMAs: Scottish Government scraps current plans on no-fishing zones following consultation. (2023). *The Scotsman*. https://www.scotsman.com/news/environment/hpmas-scottish-government-scraps-current-plans-on-no-fishing-zones-following-consultation-4201486

International Union for the Conservation of Nature (IUCN). (2020). *International cooperation on marine pollution from sunken vessels* (WCC-2020-Res-026-EN). https://portals.iucn.org/library/sites/library/files/resrecfiles/WCC_2020_RES_026_EN.pdf

Jouffray, J.-B., Blasiak, R., Norström, A. V., Österblom, H., & Nyström, M. (2020). The blue acceleration: The trajectory of human expansion into the ocean. *One Earth, 2*(1), 43–54. https://doi.org/10.1016/j.oneear.2019.12.016

Khakzad, S., & Griffith, D. (2016). The role of fishing material culture in communities' sense of place as an added-value in management of coastal areas. *Journal of Marine and Island Cultures, 5*(2), 95–117. https://doi.org/10.1016/j.imic.2016.09.002

Lloyd's Register Foundation Heritage and Education Centre. (2020, May 27). *Saving lives of the world's poorest fishermen*. Impact Stories. https://www.lrfoundation.org.uk/en/impact-stories/saving-lives-of-the-worlds-poorest-fishermen/

Lloyd's Register Foundation Heritage and Education Centre. (2021, March 24). *Bringing commercial fishermen safely home*. Impact Stories. https://www.lrfoundation.org.uk/en/impact-stories/hook-marine

Lloyd's Register Foundation Heritage and Education Centre. (2022, June 17). *Keeping fishers safe at sea*. News. https://www.lrfoundation.org.uk/en/news/keeping-fishers-safe-at-sea/

Lloyd's Register Foundation Heritage and Education Centre. (n.d.-a). *Introducing learning from the past: Drawing on our heritage to address modern issues.* Case Studies. https://hec.lrfoundation.org.uk/get-involved/introducing-learning-from-the-past-236

Lloyd's Register Foundation Heritage and Education Centre. (n.d.-b). *Ocean Decade Heritage Network.* Case Studies. https://hec.lrfoundation.org.uk/get-involved/ocean-decade-heritage-network

Majcher, J., Quinn, R., Plets, R., Coughlan, M., McGonigle, C., Sacchetti, F., & Westley, K. (2021). Spatial and temporal variability in geomorphic change at tidally influenced shipwreck sites: The use of time-lapse multibeam data for the assessment of site formation processes. *Geoarchaeology, 36*(3), 429–454. https://doi.org/10.1002/gea.21840

Marine Accident Investigation Branch. (2015). *Capsize and sinking of beam trawler Noordster with loss of 3 lives* [Marine Accident Investigation Branch reports]. https://www.gov.uk/maib-reports/capsize-and-sinking-of-beam-trawler-noordster-off-beachy-head-england-with-loss-of-3-lives

Marine Accident Investigation Branch. (2022, January 20). *Galwad-Y-Mor report published.* News Story.

McConnaughey, R. A., Hiddink, J. G., Jennings, S., Pitcher, C. R., Kaiser, M. J., Suuronen, P., Sciberras, M., Rijnsdorp, A. D., Collie, J. S., Mazor, T., Amoroso, R. O., Parma, A. M., & Hilborn, R. (2020). Choosing best practices for managing impacts of trawl fishing on seabed habitats and biota. *Fish and Fisheries, 21*(2), 319–337. https://doi.org/10.1111/faf.12431

Muckelroy, K. (1977). Historic wreck sites in Britain and their environments. *International Journal of Nautical Archaeology, 6*(1), 47–57. https://doi.org/10.1111/j.1095-9270.1977.tb00987.x

National Federation of Fishermen's Organisations. (2022, October 18). *Spatial squeeze report: Parliamentary launch.* News. https://www.nffo.org.uk/spatial-squeeze-report-parliamentary-launch/

Ndungu, K., Beylich, B. A., Staalstrøm, A., Øxnevad, S., Berge, J. A., Braaten, H. F. V., Schaanning, M., & Bergstrøm, R. (2017). Petroleum oil and mercury pollution from shipwrecks in Norwegian coastal waters. *Science of the Total Environment, 593–594*, 624–633. https://doi.org/10.1016/j.scitotenv.2017.03.213

Papageorgiou, M. (2018). Underwater cultural heritage facing maritime spatial planning: Legislative and technical issues. *Ocean & Coastal Management, 165*, 195–202. https://doi.org/10.1016/j.ocecoaman.2018.08.032

Plets, R., Quinn, R., Forsythe, W., Westley, K., Bell, T., Benetti, S., McGrath, F., & Robinson, R. (2011). Using multibeam echo-sounder data to identify shipwreck sites: Archaeological assessment of the Joint Irish Bathymetric Survey data. *International Journal of Nautical Archaeology, 40*(1), 87–98. https://doi.org/10.1111/j.1095-9270.2010.00271.x

Quinn, R., & Boland, D. (2010). The role of time-lapse bathymetric surveys in assessing morphological change at shipwreck sites. *Journal of Archaeological Science, 37*(11), 2938–2946. https://doi.org/10.1016/j.jas.2010.07.005

Quinn, R., & Smyth, T. A. G. (2018). Processes and patterns of flow, erosion, and deposition at shipwreck sites: A computational fluid dynamic simulation. *Archaeological and Anthropological Sciences, 10*(6), 1429–1442. https://doi.org/10.1007/s12520-017-0468-7

S&P Global. (2020, May 28). Seafarer fatalities shows shipping must do more. *Market Intelligence.* https://www.spglobal.com/marketintelligence/en/mi/research-analysis/seafarer-fatalities-shows-shipping-must-do-more.html

Scottish Government. (2022, December 12). *Overview.* Scottish Highly Protected Marine Areas (HPMAs). https://consult.gov.scot/marine-scotland/scottish-highly-protected-marine-areas/

SKIPINNISH. (2023, April 13). *The clearances again [Official Video].* https://www.youtube.com/watch?v=0tzW8EYDDE0

Smyth, T. A. G., & Quinn, R. (2014). The role of computational fluid dynamics in understanding shipwreck site formation processes. *Journal of Archaeological Science, 45*, 220–225. https://doi.org/10.1016/j.jas.2014.02.025

Szafrańska, M., Gil, M., & Nowak, J. (2021). Toward monitoring and estimating the size of the HFO-contaminated seabed around a shipwreck using MBES backscatter data. *Marine Pollution Bulletin, 171*, 112747. https://doi.org/10.1016/j.marpolbul.2021.112747

Turner, P. J., Cannon, S., DeLand, S., Delgado, J. P., Eltis, D., Halpin, P. N., Kanu, M. I., Sussman, C. S., Varmer, O., & Van Dover, C. L. (2020). Memorializing the middle passage on the Atlantic seabed in areas beyond national jurisdiction. *Marine Policy, 122*, 104254. https://doi.org/10.1016/j.marpol.2020.104254

UNESCO. (2001). *Convention on the protection of the underwater cultural heritage*. https://www.unesco.org/en/legal-affairs/convention-protection-underwater-cultural-heritage?hub=66535

United Nations. (2023). *Intergovernmental conference on an international legally binding instrument under the United Nations Convention on the Law of the Sea on the conservation and sustainable use of marine biological diversity of areas beyond national jurisdiction (General Assembly resolution 72/249)*. https://www.un.org/bbnj/

Westley, K., Plets, R., Quinn, R., McGonigle, C., Sacchetti, F., Dale, M., McNeary, R., & Clements, A. (2019). Optimising protocols for high-definition imaging of historic shipwrecks using multibeam echosounder. *Archaeological and Anthropological Sciences, 11*(7), 3629–3645. https://doi.org/10.1007/s12520-019-00831-6

Willis, S., Bygvraa, D. A., Hoque, M. S., Klein, E. S., Kucukyildiz, C., Westwood-Booth, J., & Holliday, E. (2023). The human cost of global fishing. *Marine Policy, 148*, 105440. https://doi.org/10.1016/j.marpol.2022.105440

Open Access This chapter is licensed under the terms of the Creative Commons Attribution 4.0 International License (http://creativecommons.org/licenses/by/4.0/), which permits use, sharing, adaptation, distribution and reproduction in any medium or format, as long as you give appropriate credit to the original author(s) and the source, provide a link to the Creative Commons license and indicate if changes were made.

The images or other third party material in this chapter are included in the chapter's Creative Commons license, unless indicated otherwise in a credit line to the material. If material is not included in the chapter's Creative Commons license and your intended use is not permitted by statutory regulation or exceeds the permitted use, you will need to obtain permission directly from the copyright holder.

Chapter 8
Beneath a Façade: The Unscientific Justification of Treasure Salvage

Michael L. Brennan

Professor Thijs Maarleveld of the University of Southern Denmark and co-founder of the ICOMOS International Committee on the Underwater Cultural Heritage (ICUCH), was a fierce guardian of underwater cultural heritage and proponent of the UNESCO recommendation of *in situ* preservation as a first option. Despite his unfortunate and untimely passing in 2020, Prof. Maarleveld's efforts to combat the misrepresentation of commercial salvage as archaeological investigation remains a beacon in the field and a reminder of best practices for protecting UCH. He championed the guiding principle that 'the preservation *in situ* of underwater cultural heritage shall be considered as the first option'. Maarleveld noted that 'the principle to consider *in situ* preservation first is not to be confused with the foregone conclusion that *in situ* preservation is what is to be decided to' (2016, p. 478), only that it is considered as a first option. In addition, strides have been made by UNESCO and other organisations to develop tools for preservation *in situ* of shipwreck sites that contribute to long term management options but do not 'reduce a central principle to a management tool' and it is even understood that there are situations and conditions in which *in situ* preservation is not a preferred or feasible option (Maarleveld, 2016). However, there are those that have used this as a way to argue instead for commercial salvage of any shipwrecks, not just those that are threatened.

If one does an online search for information about trawl damage to shipwreck sites, a number of grey literature sources come up relating to the work conducted by treasure hunting company Odyssey Marine Exploration, which attempts to exploit the fact that some shipwrecks are damaged by bottom trawling activities into a broad justification for commercial salvage. Through their work surveying in the English Channel, upon observing that many shipwrecks there exhibited evidence of impacts from fishing gear, the company used this as an opportunity to argue against

M. L. Brennan (✉)
SEARCH Inc., Jacksonville, FL, USA
e-mail: mike@brennanexploration.com

UNESCO's principle of *in situ* preservation as the first option. However, the recovery of entire shipwrecks-worth of artefacts is not the answer to the problem of trawl damage to wrecks, as this volume addresses. What is needed is further documentation of trawl damage to shipwrecks, which in turn requires expanded deep-water exploration to locate, document, and conduct repeat visits in order to quantify damage (Brennan et al., 2016). Most imperative is locating sites in deep water within range of trawls, particularly depths from 200 to 2000 m. It is as important to protect known shipwrecks from trawling as it is to find unknown wrecks to document and protect, as mentioned in a previous chapter by Brennan (Chap. 4, this volume). Expanded funding for exploration is sorely needed. We cannot protect or manage shipwreck sites we don't know about.

That said, the publications by Odyssey Marine are professionally formatted with the appearance of an academic publication. The flashy logo and colour photos engage the reader with an appearance of scientific rigor, which is only present on the surface, and which archaeologists have referred to as creating 'an illusion of research' (Greene et al., 2011, p. 115). Their Oceans Odyssey books and handful of other papers published on their website have no overt reference to them having been peer reviewed, which is troubling. Why is this important? Peer review is done by external reviewers with familiarity of the subject to ensure the work is original, scholarly and in the case of this subject, meets professional archaeological standards including the use of appropriate scientific methodologies, and that the projects were conducted in accordance with professional ethics of archaeologists and conservators. Odyssey Marine used glossy underwater photographs of damage to shipwrecks from bottom-fishing activities to justify the 'salvage' of our common heritage and sale of it for private profit. It of course, does not save our heritage for future generations and instead uses public resources for personal gain. In all of Odyssey's writings, there is a clear, systematic justification of their salvage being presented to the reader.

Four volumes of 'Oceans Odyssey' were released through the Odyssey Marine website and printed by Oxbow Books. They contain a variety of papers by expedition personnel that strive to look like an assemblage of scholarly works. In the Preface, Odyssey Marine founder, Greg Stemm, lays out the idea that trawl damage to wrecks justifies their commercial salvage and sale: 'We found that even in deep water, shipwrecks were being destroyed at an alarming rate and that the politics of underwater cultural heritage were so complex that some government bureaucrats were happier to see shipwrecks being destroyed *in situ* than to consider a new private sector model for managing cultural heritage… It was a very expensive endeavour – far beyond the budgets of academic archaeological institution… we also allowed for generating of profits by… making large quantities of duplicate artefacts – such as coins – available to the public' (Stemm, 2010, p. vii).

Their plan for the privatisation of our public heritage is further presented by John Kimball in Oceans Odyssey 2: 'Only artefacts that fit our Trade Good definition are offered for sale. This is a category characterized by large quantities of mass-produced objects, such a coins, bottles, pottery and other mass-produced cargo… Duplicates are only sold to private collectors after thorough study and recording'

(2011, p. 14). This business model is 'commercial exploitation' and inconsistent with the 2001 Convention and US historic preservation law (Ole Varmer, personal communication). The fact that some shipwreck sites are threatened by anthropogenic or natural factors is convenient to their objective. However, it does not overcome the conflicts with heritage law and policy or justify privatising public resources. Kimball further writes: 'Odyssey has discovered hundreds of shipwrecks... Our policy is to record the site, and then either pick up a small selection of diagnostic artefacts for study and permanent retention or, in the majority of cases, leave the site undisturbed *in situ*' (2011, p. 16), which further illustrates that they only recover things of monetary value that they can sell for a profit. A major problem with Odyssey Marine's business model is the speculative sale of duplicate coins and other artefacts to try and secure funds for exploration and recovery. As indicated, there is also the problem of the sale of artefacts salvaged from shipwrecks resulting in the privatisation of public resources. Such sales then may be used to raise funds for future treasure hunting, commercial salvaging, if not looting, of other wrecks. In reference to the Oceans Odyssey volumes, Liz Greene writes, 'such seemingly innocuous descriptions serve as a veiled justification for the sale of artefacts and reflect Odyssey CEO Greg Stemm's desire to separate 'cultural artefacts' and 'trade goods' so the latter can be sold on the open market' (Greene et al., 2011, p. 115).

Stemm leans further into the façade of good science in the Oceans Odyssey I Preface regarding archaeologists' views: 'Their stated position is that any company with a profit motive could not possibly be concerned with science' (Stemm, 2010, p. viii). This is far from the case, as cultural resource management (CRM) firms conduct rigorous archaeology worldwide, particularly in the United States, where it is reported to, reviewed by, and approved by both State and Federal archaeologists for concurrence. Numerous government and private industry archaeological projects have developed into peer reviewed publications, for example that on USS *Monitor* (Broadwater, 2012), Japanese midget submarines off Pearl Harbor (Delgado et al., 2016), World War II shipwrecks in the Gulf of Mexico (Church & Warren, 2008), and the wreck of *Clotilda* in the Mobile River (Delgado et al., 2023), just to name a few. If Odyssey Marine was so concerned with science, where are their academic publications? What journal articles have come out using any of the scientific data collected by any of these expeditions? In one chapter, Sean Kingsley emphasises the observation that there are a scarcity of shipwrecks predating 1800 in the English Channel (Kingsley, 2010, p. 226). The implication here is that bottom trawling has erased them entirely. It is possible this occurred. However, if a scientific approach were to be taken in writing this chapter, sedimentation, sediment load from nearby rivers, burial, tidal flux and current dynamics would be researched and presented, but no such information is even hinted at. Such omissions would likely have been caught by peer reviewers. Odyssey Marine's publications are an attempt to appear legitimate and scientific, only to then rationalise commercial sales of their recovered artefacts. However, not only is the work unscientific, but the work is also unauthorised and unregulated, and often destructive.

Kingsley's later book, *Fishing and Shipwreck Heritage*, is not officially part of the Oceans Odyssey series but has the same underlying objective. In the Preface, Kingsley complains that the UK's Joint Nautical Archaeology Policy Committee (JNAPC) tried to 'suppress and discredit all research' by Odyssey Marine in an effort to stop them from excavating the wreck of HMS *Victory*, which Odyssey Marine discovered in 2008 (Kingsley, 2016, p. xiv). The author states that the committee questioned the evidence of trawling presented by Odyssey Marine. While that may be the case, it is clear that the JNAPC did not want to see Odyssey Marine involved in the recovery of British heritage particularly the plan that the artefacts were to be sold. This led to statements such as 'the notion of selling shipwreck artefacts, from potsherds to coins, whether for direct profit or to channel back into funding science, is portrayed as the slaying of archaeology's most sacred cow' (2016, p. 9). Such actions are what led to the UNESCO Annex Rule 2 that 'the commercial exploitation of underwater cultural heritage for trade or speculation or its irretrievable dispersal is fundamentally incompatible with the protection and proper management of underwater cultural heritage. Underwater cultural heritage shall not be traded, sold, bought or bartered as commercial goods' (as cited by Kingsley, 2016, p. 10). Bottom trawl fishing is a problem for the preservation of shipwrecks in many parts of the world, and as this volume illustrates, it is one the archaeological community is working to address. Blatant and intentional destruction of a shipwreck site for the purpose of selling the artefacts for profit, regardless of whatever 'science' is done ahead of it, goes against all principles of archaeology. Kingsley also states that 'wrecks subjected to treasure hunting are typically not published scientifically' (2016, p. 11). While trying to separate Odyssey Marine from other treasure hunters, as discussed previously, Odyssey Marine's publications do not qualify as scientific.

In his review of Kingsley's book, Prof. Maarleveld wrote, 'neither fishing nor heritage lie at the heart of this book. Rather, I get the impression that they serve as a decoy for continuous complaint about the archaeological profession, authorities, international organizations such as UNESCO, regulations, committees and bureaucracy that curb the freedom of action of everyone, but first and foremost of Dr Kingsley himself' (Maarleveld, 2016, p. 478). In other words, Odyssey Marine's arguments regarding trawling damage are not so much that archaeological sites are being damaged, but if fishers can damage sites, why can't we? 'Much treasure hunting has disguised itself with an image of respectable explorative research' Maarleveld further wrote (2016, p. 479). The hidden agenda of Odyssey Marine is not a new tactic. Maarleveld concludes his review by directing readers to another book in the same series as Kingsley's by Colin Renfrew, *Loot, Legitimacy and Ownership*, which 'explains why archaeologists should steer clear of operations purporting to work as archaeologists while selling artefacts and promoting the market for archaeological objects' (p. 479).

Mischaracterisations plague Odyssey Marine's writings as they try to warp actual scientific publications to fit their narrative. One example is some of the work I've put forth from expeditions in the Black Sea. *Nautilus* expeditions in the Black Sea in 2011 and 2012 followed work by Robert Ballard in 2000 and 2003 that located four Byzantine shipwrecks off Sinop. These expeditions discovered additional shipwrecks, both off Sinop as well as off Ereğli along the northern coast of Turkey (Brennan et al., 2013, 2016). Many of these wrecks exhibited trawl damage, which we had also documented on ancient shipwrecks in the Aegean Sea (Brennan et al., 2012, Brennan, Chap. 4, this volume). Like off southwestern Turkey in the Aegean, we proposed marine protected areas and regions of additional trawling prohibitions to begin to protect these newly discovered wrecks (Brennan et al., 2012; Krumholz & Brennan, 2015). We concluded, 'It is essential to continue to conduct rapid comprehensive surveys of such threatened areas before these sites are damaged further, potentially beyond the point at which they can be detected' (Brennan et al., 2012, p. 69, cited in Kingsley, 2016, p. 95). Kingsley states that 'The team's deductions are hard to square with their own additional conclusions' (p. 95). The point made in our Black Sea articles is the same as here: we cannot protect underwater cultural resources we do not know about. Ocean exploration, especially in depths that are in the range of trawlers (200–2000 m), is essential to locate wrecks that need to be protected through marine protected areas (MPAs) or other exclusion zones. This is a consistent argument throughout all the publications related to trawling stemming from the *Nautilus* expeditions.

One wreck in the Black Sea, Ereğli E, was of particular importance for a number of reasons. It is the oldest shipwreck in deep water found along the southern Black Sea coast of Turkey, and also exhibited extensive damage from trawls (Brennan et al., 2013, 2016; Davis et al., 2018). The wreck was discovered in 2011 and mapped with video, still cameras and multibeam sonar. In this imagery were objects that we suspected may be human bones, which were likely trawled up from below the mudline. Upon our return and remapping of the site in 2012, approximately 11 months later, the site had been further trawled, many of the artefacts moved offsite, and the bones were no longer visible (Brennan et al., 2016). Kingsley states 'This data loss can hardly have come as a shock' (2019, p. 18). He adds, 'leaving wrecks *in situ* without selective sampling or excavation seems at best counterintuitive' (2016, p. 96). Odyssey Marine may not have been too concerned about obtaining permits. *Nautilus* was operating in Turkish waters and had Turkish observers aboard with specific stipulations in our marine scientific research permits through the Ministry of Foreign Affairs that prohibited touching or removing anything from a shipwreck. We were very aware of the fragile state of Ereğli E and would have collected the bones for analysis had we been able. In fact, when we returned in 2012, it was with specific permission from Turkey to recover the bones that we sought through proper channels. It was an unfortunate circumstance that they were no longer there. However, the urgency of a site's imminent danger from trawls does not supersede the need for legal compliance through permissions from the coastal state.

In 2007, Odyssey Marine located the wreck of what they called the 'Black Swan' that was subsequently proven to be the Spanish Navy frigate, *Nuestra Señora de las Mercedes*. *Mercedes* sank off Portugal in 1804 with a cargo of gold and silver specie in a battle with the British. Evidence showed Odyssey Marine's discovery came after researching the potential location of the wreck in order to find and recover its treasure. After doing so, they moved the artefacts to Gibraltar for a quick air shipment home to Tampa and the filing of a claim in a Federal Admiralty court. Spain intervened in the US court to contest the Florida-based company's assertion of rights to recover the artefacts taken from the wreck site on the continental shelf of Portugal, which remained the property of the State and subject to sovereign immunity from an admiralty arrest since it was a sunken state craft. Odyssey Marine claimed in court that the 17 tonnes of coins it had recovered were jettisoned cargo from an unknown ship (Delgado & Goold, 2021, p. 352). Video and photographic data were mischaracterised, diagnostic artefacts either hidden in a locked storage unit in Gibraltar or left undocumented on the site, and knowingly false statements were made to the court. While their operations were self-proclaimed to be scientific and adhering to archaeological standards, what the court unveiled was that recovery operations by Odyssey Marine were 'consistent with a sustained effort using sophisticated ROV systems to recover as many of the 900,000 pesos presumed to be on the site as could be accomplished in a one-two month period… [and] very few artefacts of other types were recovered' (Delgado & Goold, 2021, p. 358). Artefacts that were recovered that were diagnostic and refuted Odyssey Marine's claim were not disclosed, including uniform buttons of a Captain of the Spanish Royal Navy Marine. If diagnostic information was collected from other artefacts, such as bronze culverins that had the Spanish coat of arms, it was not disclosed, and Spain later documented these during their own visit to the site. As a witness in the case, Kingsley stated under oath: '[n]o stamped or incised epigraphic evidence has been recorded… to identify the Black Swan site' (Delgado & Goold, 2021, p. 359). In other words, Odyssey Marine intentionally did not record essential data for proper archaeology.

The Federal Admiralty court in Florida noted that Odyssey Marine researched *Mercedes* and intentionally surveyed waters off Portugal for the wreck, that the wreck's identity was 'well known to Odyssey from the start', and that their efforts had 'concentrated on a sustained effort to recover coins and little else' (Delgado & Goold, 2021, p. 359). That is not scientific. That is not archaeology. That is commercial salvage and treasure hunting. The court ultimately ordered the coins returned to Spain and fined Odyssey Marine in excess of $1 million for false representations to the court and to Spain. This case serves as a prime example of how this for-profit company was not concerned with legitimate archaeological or scientific practices, despite what they may try to market or project.

Returning to the subject at hand of bottom trawl fishing damage to shipwrecks, Kingsley undercuts his entire argument in the conclusion of his book: 'When confronted by the figure of three million shipwrecks worldwide, this aspiration appears not just daunting but crippling. The number of sites actually requiring attention, however, is a minimal percentage of the global total' (Kingsley, 2016, p. 118). It is clear that his concern for shipwrecks being damaged does not apply equally to all shipwrecks. Ostensibly, those of interest are those of potential commercial value.

Odyssey found 267 wrecks in the English Channel, and they concluded that only a handful of those deserved full excavation—likely to be followed by recovery and sale of artefacts. What about the rest? It hardly has to be said that it is not possible to fully excavate and recover every shipwreck in the world. This is one of the main reasons behind the *in situ* preservation policy of professional archaeologists, and codified in domestic and international law including the UNESCO 2001 Convention. It is not a prohibition on recovery under the Annex rules but rather a policy or practice 'as the first option' to be considered in the management of the resource in the public interest and not for private profit. Who should decide what wreck is most important to excavate, and based on what? Its potential value to collectors? What about the other 200+ wrecks in the English Channel? Should those deemed less important (i.e., less valuable) be left to be destroyed? There are solutions that can help protect all shipwrecks, those deemed valuable and those mundane wrecks that treasure hunters would not care about, but which hold the stories of everyday mariners that are equally engaging and historically important.

Potential solutions become apparent with advances in technology, such as satellite and digital infrastructure, that can enable government and international organisational oversight of known shipwreck sites, be it in established Marine Protected Areas or some other sort of exclusion zone. This can include all types of shipwrecks, even those deemed mundane. Kingsley (2016) states: 'vast slices of the world's sunken history are almost impossible to police' (p. 120) and considers such efforts 'a fool's paradise' (2012, p. 24). This is convenient for his argument, and also incorrect. Through tracking of vessels by Automatic Identification Systems (AIS) and other global positional systems, management of sensitive areas and sites is achievable with the right level of funding and infrastructure, and we have been moving in that direction.

A good example is Stellwagen National Marine Sanctuary located offshore of New England. The Shipwreck Avoidance Pilot Program is making an effort to protect shipwrecks within the sanctuary, which does not prohibit fishing from active trawlers. The program disclosed wreck locations and installed geofences around them (Trethewey, 2023; Mires et al., Chap. 6, this volume). When a vessel crossed one of these boundaries, monitored through its GPS navigation system, a warning would come up stating, 'Captain, your vessel has entered a shipwreck avoidance area… NOAA requests that you keep your gear at least 400 feet [122 m] away' (Trethewey, 2023). This system is a strong stride forward. A colleague and I proposed a series of marine protected areas (MPAs) for the wrecks *Nautilus* found off Turkey in high densities in the Black and Aegean Seas over a series of expeditions (Krumholz & Brennan, 2015). The argument is that protection of such resources, which serve as hard substrate and artificial reefs, can allow for juvenile fish to survive and 'spillover' into fishable areas, thereby both protecting shipwrecks and helping to sustain local fisheries. Such marine protected areas do not, however, need to be wide swaths of seabed in the traditional MPA structure, but could be as small as geofences that modern technology can assist with ensuring fisher avoidance.

Recent international treaties have continued to further the implementation of protected areas worldwide. The United Nations' Biodiversity Beyond National Jurisdiction (BBNJ) resolution is a new legally binding instrument adopted in June 2023. This treaty addresses, among other things, Area Based Management Tools (ABMT) and marine protected areas and includes protective measures that all ships must adhere to (IMO, 2023). Such sensitive areas could include a broader use of geofences around areas of known shipwreck sites, noting that such geofences can be larger areas so as not to completely disclose a wreck's location. I propose that an international automated system be established with stronger penalties for trawlers entering shipwreck geofences and MPAs or turning off their AIS, including fines that can be automatically deducted from an account, similar to EZPass for tolls on the highways in the United States. In today's era of modern technology, this is realistic. What is required, in addition to funding and organising this sort of infrastructure development and implementation, is vastly larger support for ocean exploration to find shipwrecks that can then be documented, managed, and ultimately, protected.

In situ preservation efforts as a first option to management of UCH does not imply that such effort will be successful in preserving a shipwreck forever. A shipwreck has been stated to have a 'life cycle' (e.g., Muckelroy, 1978; Brennan et al., 2011) that begins with the vessel's sinking until which time it is either degraded or corroded to the point that it no longer exists, or it is completely buried in sediment and no longer discoverable. Such a life cycle is natural. Protection of shipwreck sites, and their preservation *in situ,* does not seek to stop this natural process. *Titanic* will one day—centuries from now—be a pile of rust on the seabed of the Atlantic Ocean; this is a fitting end to the site as the resting place of 1,517 souls. A shipwreck is like an ancient redwood tree in the forest that will one day die and fall. *In situ* preservation efforts, archaeological characterisation of shipwreck sites, and the high-resolution documentation of them—which increases as technology advances— is the work of maritime archaeologists to preserve the shipwreck is as much detail for the historical record as possible before this happens. In the same way that we would make every effort to protect the redwood from forest fires, we would also protect it from those looking to cut the tree down to profit from its wood. That is the effort by UNESCO and archaeologists to prevent both bottom trawlers and treasure hunters from destroying shipwreck sites, whether incidentally or for profit. Similarly, while the warships sunk in World Wars I and II will corrode and deteriorate over time, that is different than the commercial salvage of these hulls for steel that eradicate the site from the seabed, which has been ongoing in the South Pacific. The fact that a shipwreck may one day disappear through natural degradation or burial does not mean it should be ripped from the seabed and its parts sold to the highest bidder.

Bottom trawling is a profound threat to our Ocean Heritage, but recent scientific work to address this threat and the implementation of modern technology can begin to establish protections for shipwrecks in jeopardy from this activity. It is certainly not a justification for the commercial salvage of valuable artefacts from select shipwrecks for sale. Nor is trawling the greatest threat to shipwrecks. As Professor Maarleveld stated in his review of *Fishing and Shipwreck Heritage*, 'the greatest threat to archaeology, is bad archaeology'.

References

Brennan, M. L., Ballard, R. D., Croff Bell, K. L., & Piechota, D. (2011). Archaeological oceanography and environmental characterization of shipwrecks in the Black Sea. In I. Buynevich, V. Yanko-Hombach, A. Gilbert, & R. E. Martin (Eds.), *Geology and geoarchaeology of the Black Sea region: Beyond the flood hypothesis* (Geological Society of America special paper 473) (pp. 179–188).

Brennan, M. L., Ballard, R. D., Roman, C., Croff, K. L., Bell, B. B., Coleman, D. F., Inglis, G., Koyagasioglu, O., & Turanli, T. (2012). Evaluation of the modern submarine landscape off southwestern Turkey through the documentation of ancient shipwreck sites. *Continental Shelf Research, 43*, 55–70.

Brennan, M. L., Davis, D., Roman, C., Buynevich, I., Catsambis, A., Kofahl, M., Derya Ürkmez, J., Vaughn, I., Merrigan, M., & Duman, M. (2013). Ocean dynamics and anthropogenic impacts along the southern Black Sea shelf examined by the preservation of pre-modern shipwrecks. *Continental Shelf Research, 53*, 89–101.

Brennan, M. L., Davis, D., Ballard, R. D., Trembanis, A. C., Ian Vaughn, J., Krumholz, J. S., Delgado, J. P., Roman, C. N., Smart, C., Bell, K. L. C., Duman, M., & DuVal, C. (2016). Quantification of bottom trawl fishing damage to ancient shipwreck sites. *Marine Geology, 371*, 82–88.

Broadwater, J. (2012). *USS monitor: A historic ship completes its final voyage*. Texas A&M University Press, College Station.

Church, R. A., & Warren, D. J. (2008). The 2004 deepwrecks project: Analysis of World War II era shipwrecks in the Gulf of Mexico. *International Journal of Nautical Archaeology, 12*(2), 82–102.

Davis, D., Brennan, M. L., Opaiţ, A., & Beatrice, J. S. (2018). The Eregli E shipwreck: A Hellenistic merchant ship in the Black Sea. *International Journal of Nautical Archaeology, 41*, 57–80.

Delgado, J. P., & Goold, J. A. (2021). Background to the 'Black Swan' case: The identification as *Nuestra Señora de las Mercedes*. *International Journal of Nautical Archaeology, 50*(2), 352–360.

Delgado, J. P., Kirby, T., Van Tilburg, H. K., Price, S., Varmer, O., Cremer, M. D., & Matthews, R. (2016). *The lost submarines of Pearl Harbor: The rediscovery and archaeology of Japan's top-secret midget submarines of World War II*. Texas A&M University Press, College Station.

Delgado, J. P., Marx, D. E., Lent, K., Grinnan, J., & DeCaro, A. (2023). *Clotilda: The history and archaeology of the last slave ship*. Texas A&M University Press, College Station.

Greene, E. S., Leidwanger, J., Leventhal, R. M., & Daniels, B. I. (2011). Mare nostrum? Ethics and archaeology in Mediterranean waters. *American Journal of Archaeology, 115*, 311–319.

International Maritime Organization. (2023, June 19). *IMO welcomes adoption of new oceans treaty*. International Maritime Organization. Available https://www.imo.org/en/MediaCentre/PressBriefings/Pages/BBNJ-treaty-adoption.aspx. Accessed 10 Sept 2023.

Kimball, J. (2011). Living with the convention on the protection of the underwater cultural heritage: New jurisdictions. In *Oceans Odyssey 2: Underwater heritage management & deep-sea shipwrecks in the English Channel & Atlantic Ocean* (Odyssey marine exploration reports 2) (pp. 12–17). Oxbow Books.

Kingsley, S. A. (2010). Deep-sea fishing impacts on the shipwrecks of the English Channel & Western approaches. In *Oceans Odyssey: Deep-sea shipwrecks in the English Channel, Straits of Gibraltar & Atlantic Ocean* (Odyssey Marine Exploration reports 1) (pp. 191–233). Oxbow Books.

Kingsley, S. A. (2012). *Out of sight, out of mind? Fishing & shipwrecked heritage*. Wreck Watch International, London, 33 p.

Kingsley, S. A. (2016). *Fishing and shipwreck heritage: Marine archaeology's greatest threat?* Bloomsbury Academic.

Kingsley, S. A. (2019). *Shipwrecked in situ: Saving the sunken past or scapegoat archaeology?* Maritime Heritage Foundation.

Krumholz, J. S., & Brennan, M. L. (2015). Fishing for common ground: Investigations of the impact of trawling on ancient shipwreck sites uncovers a potential for management synergy. *Marine Policy, 61*, 127–133.

Maarleveld, T. J. (2016). Book review Fishing and shipwreck heritage: Marine archaeology's greatest threat? *International Journal of Nautical Archaeology, 45*(2), 478–479.

Muckelroy, K. (1978). *Maritime archaeology*. Cambridge University Press.

Stemm, G. (2010). Preface. In *Oceans Odyssey: Deep-sea shipwrecks in the English Channel, Straits of Gibraltar & Atlantic Ocean* (Odyssey marine exploration reports 1) (pp. vii–viii). Oxbow Books, Oxford.

Trethewey, L. (2023, August 22). Loose lips save sunken ships. *Hakai Magazine*. Available at https://hakaimagazine.com/features/loose-lips-save-sunken-ships/. Accessed 10 Sept 2023.

Open Access This chapter is licensed under the terms of the Creative Commons Attribution 4.0 International License (http://creativecommons.org/licenses/by/4.0/), which permits use, sharing, adaptation, distribution and reproduction in any medium or format, as long as you give appropriate credit to the original author(s) and the source, provide a link to the Creative Commons license and indicate if changes were made.

The images or other third party material in this chapter are included in the chapter's Creative Commons license, unless indicated otherwise in a credit line to the material. If material is not included in the chapter's Creative Commons license and your intended use is not permitted by statutory regulation or exceeds the permitted use, you will need to obtain permission directly from the copyright holder.

Chapter 9
Conclusion: Looking Forward

Charlotte Jarvis

The case studies presented in this volume offer a compelling look at the damage caused to many forms of Underwater Cultural Heritage (UCH) by bottom trawling and other mobile fishing gear, as well as provide some suggestions to protect this vital Ocean Heritage resource for future generations. The authors from Stellwagen Bank National Marine Sanctuary (Chap. 6, this volume) highlight their new Sanctuary Mapping Initiative which works with fishers to 'to conduct side-scan sonar surveys to locate and document shipwrecks and characterise seafloor habitats'. It is a promising step forward and can help to show fishers in other waters the shared natural and cultural importance to shipwrecks and highlight the need for protection of seabed heritage.

Michael Brennan's work in Turkey indicates this is possible: his data shows 'curves to the west as fishers began veering away while they recovered their towed gear', indicating 'that the trawlers in this coastal area abided by local regulations' (Brennan, Chap. 4, this volume). A desire to reach this sort of compliance is sorely needed, as noted in Chap. 7, that focuses on trawling and fishing communities. The authors call for more effort 'to develop ways to balance the needs of fishers with preservation of nature and heritage' (this volume).

New technology allows this balance to be monitored in new ways. Brennan highlights throughout his work in this volume the potential use of AIS monitoring. AIS is carried by most vessels, so active trawlers can be monitored. This can 'provide the resources necessary to ensure that deep-water shipwrecks are no longer out of sight and out of mind' (Brennan, Chap. 4, this volume). Brennan argues for a type of EZPass (as used for tolls on the highways in the United States) to implement this. Geofences can be utilised to ensure that when a trawler crosses into, for example, areas of high numbers of shipwrecks or a vessel switches off its AIS, a fine is levied.

C. Jarvis (✉)
The Ocean Foundation, Washington, DC, USA

What will be most beneficial in accomplishing these solutions is an increase in monitoring shipwrecks at risk from trawling and a greater pool of case studies to learn from. As noted by Majcher et al. in Chap. 5, 'advancements in geophysical techniques have afforded researchers unprecedented insights into the dynamics of underwater shipwreck sites'. Project ENDURE is working towards using new insights to increase the amount of data and analysis available for scientists to monitor shipwrecks at risk from bottom trawling. The archaeological data should also be considered alongside natural ocean science data, for example the Census of Marine Life,[1] to see the impacts on both. Trawling threatens not just UCH, but also natural resources in the ocean.

This damage from bottom trawling also feeds into another volume in this series, *Threats to Our Ocean Heritage: Potentially Polluting Wrecks* (2024). Here, editor Michael Brennan pulls together case studies that demonstrate the risks to Ocean Heritage from Potentially Polluting Wrecks (PPWs). In several cases, wrecks surveyed as PPWs have also showed evidence of bottom trawling impacts and it is possible that those impacts contributed to an oil leak (Brennan et al., 2023).

9.1 Maritime Heritage Ecology and Steps Forward

UCH can support ecological marine biodiversity and helps boost sea connectivity. Kirstin S. Meyer-Kaiser and Calvin H. Mires coined the term 'Maritime Heritage Ecology', a field that must be further explored (Meyer-Kaiser & Mires, 2022) and the importance of which is highlighted in Chap. 6. Cultural heritage and natural heritage are intertwined: UCH can support ecological marine biodiversity and help boost sea connectivity. Natali Pearson and Benjamin Thompson argue that it is beneficial for sites with high UCH and high natural heritage to co-occur and be used strategically together (Pearson & Thompson, 2023).

Finally, damage from bottom trawling provides a look into the potential future of our Ocean Heritage if effective regulations are not passed for Deep Seabed Mining (DSM). What we do with trawling can pave the way for other ocean exploration and exploitation (Jarvis, 2023). Additionally, while there are many unknowns in deep sea biodiversity and ecosystem services, what is already known clearly points to vast and far-reaching damage by humans. In other words, we know enough from existing trawling damage that tells us we should stop similar practices, like seabed mining, and look to regulate existing practices. Trawling is a clear and present threat not only to fish populations and habitats, but also to submerged cultural landscapes. There should not be a choice between humans and the natural world. Restrictions, such as the options outlined in this volume, must be passed as soon as possible to prevent further material and heritage destruction.

---

[1] http://www.coml.org/index.html

References

Brennan, M. L., Delgado, J. P., Jozsef, A., Marx, D. E., & Bierwagen, M. (2023). Site formation processes and pollution risk mitigation of World War II oil tanker shipwrecks: Coimbra and Munger T. Ball. *Journal of Maritime Archaeology, 18*(2), 321–335. https://doi.org/10.1007/s11457-023-09365-4

Jarvis, C. (2023, November). Bottom trawling and the damage to underwater cultural heritage: An overview of the destruction and possible steps forward. In *Asia Pacific regional conference on underwater cultural heritage proceedings. 5th Asia Pacific regional conference on underwater cultural heritage*, Gwangju, South Korea.

Meyer-Kaiser, K. S., & Mires, C. H. (2022). Underwater cultural heritage is integral to marine ecosystems. *Trends in Ecology & Evolution, 37*(10), 815–818. https://doi.org/10.1016/j.tree.2022.06.014

Pearson, N., & Thompson, B. S. (2023). Saving two fish with one wreck: Maximizing synergies in marine biodiversity conservation and underwater cultural heritage protection. *Marine Policy, 152*, 105613. https://doi.org/10.1016/j.marpol.2023.105613

Open Access This chapter is licensed under the terms of the Creative Commons Attribution 4.0 International License (http://creativecommons.org/licenses/by/4.0/), which permits use, sharing, adaptation, distribution and reproduction in any medium or format, as long as you give appropriate credit to the original author(s) and the source, provide a link to the Creative Commons license and indicate if changes were made.

The images or other third party material in this chapter are included in the chapter's Creative Commons license, unless indicated otherwise in a credit line to the material. If material is not included in the chapter's Creative Commons license and your intended use is not permitted by statutory regulation or exceeds the permitted use, you will need to obtain permission directly from the copyright holder.

Printed by Printforce, the Netherlands